Science and Technology Concepts–Secondary™

Understanding
Weather
and Climate

Student Guide

National Science Resources Center

The National Science Resources Center (NSRC) is operated by the Smithsonian Institution to improve the teaching of science in the nation's schools. The NSRC disseminates information about exemplary teaching resources, develops curriculum materials, and conducts outreach programs of leadership development and technical assistance to help school districts implement inquiry-centered science programs.

Smithsonian Institution

The Smithsonian Institution was created by an act of Congress in 1846 "for the increase and diffusion of knowledge..." This independent federal establishment is the world's largest museum complex and is responsible for public and scholarly activities, exhibitions, and research projects nationwide and overseas. Among the objectives of the Smithsonian is the application of its unique resources to enhance elementary and secondary education.

STC Program™ Project Sponsors

National Science Foundation

Bristol-Meyers Squibb Foundation

Dow Chemical Company

DuPont Company

Hewlett-Packard Company

The Robert Wood Johnson Foundation

Carolina Biological Supply Company

Science and Technology Concepts–Secondary™

Understanding
Weather
and Climate

Student Guide

The STC Program™

Smithsonian Institution
National Science Resources Center

www.carolinacurriculum.com

Published by Carolina Biological Supply Company
Burlington, North Carolina

NOTICE This material is based upon work supported by the National Science Foundation under Grant No. ESI-9618091. Any opinions, findings, and conclusions or recommendations expressed in this material are those of the authors and do not necessarily reflect views of the National Science Foundation or the Smithsonian Institution.

This project was supported, in part, by the **National Science Foundation**.
Opinions expressed are those of the authors and not necessarily those of the foundation.

ISBN 978-1-4350-0655-3

Published by Carolina Biological Supply Company, 2700 York Road, Burlington, NC 27215.
Call toll free 1-800-334-5551.

1306

Science and Technology Concepts–Secondary™ *Understanding Weather and Climate*

The following revision was based on the STC/MS™ module *Catastrophic Events*.

Developers
Amy Charles
Kitty Lou Smith

Scientific Reviewer
Amanda Babson
Environmental Protection Agency

Ian MacGregor
Senior Scientist
National Science Resources Center

Arnold Gruber
National Oceanic and Atmospheric
Administration

Illustrator
John Norton

Writers/Editors
Amy Charles
Ian Mark Brooks
Devin Reese

Photo Research
Jane Martin
Devin Reese

National Science Resources Center Staff

Executive Director
Thomas Emrick

Program Specialist/Revision Manager
Elizabeth Klemick

Contractor, Curriculum Research and Development
Devin Reese

Publications Graphics Specialist
Heidi M. Kupke

Carolina Biological Supply Company Staff

Director of Product and Development
Cindy Morgan

Marketing Manager, STC–Secondary™
Jeff Frates

Curriculum Editors
Lauren Eggiman
Gary Metheny

Managing Editor, Curriculum Materials
Cindy Vines Bright

Publications Designers
Trey Foster
Charles Thacker
Weldon D. Washington II
Greg Willette

Science and Technology Concepts for Middle Schools™
Catastrophic Events
Original Publication

Module Development Staff

Developer/Writer
Carol O'Donnell

Science Advisors

Stan Doore, Meteorologist (retired)
National Weather Service
National Oceanic and Atmospheric
Administration

Ann Dorr, Earth Science Teacher (retired)
Fairfax County Public Schools, Virginia
Board Member, Minerals Information Institute

Ian MacGregor, Director
Division of Earth Sciences
National Science Foundation

Grant Woodwell, Professor of Geology
Mary Washington College

Thomas Wright, Geologist
National Museum of Natural History
Smithsonian Institution
U.S. Geological Survey (emeritus)

Contributing Writer
Elaine Friebele

Illustrators
John Norton
Max-Karl Winkler

STC/MS™ Project Staff

Principal Investigator
Douglas Lapp, Executive Director, NSRC
Sally Goetz Shuler, Deputy Director, NSRC

Project Director
Kitty Lou Smith

Curriculum Developers
David Marsland
Henry Milne
Carol O'Donnell
Dane J. Toler

Illustration Coordinator
Max-Karl Winkler

Photo Editor
Janice Campion

Graphic Designer
Heidi M. Kupke

STC/MS™ Project Advisors

Judy Barille, Chemistry Teacher, Fairfax County Public Schools, Virginia

Steve Christiansen, Science Instructional Specialist, Montgomery County Public Schools, Maryland

John Collette, Director of Scientific Affairs (retired), DuPont Company

Cristine Creange, Biology Teacher, Fairfax County Public Schools, Virginia

Robert DeHaan, Professor of Physiology, Emory University Medical School

Stan Doore, Meteorologist (retired), National Weather Service, National Oceanic and Atmospheric Administration

Ann Dorr, Earth Science Teacher (retired), Fairfax County Public Schools, Virginia; Board Member, Minerals Information Institute

Yvonne Forsberg, Physiologist, Howard Hughes Medical Center

John Gastineau, Physics Consultant, Vernier Corporation

Patricia A. Hagan, Science Project Specialist, Montgomery County Public Schools, Maryland

Alfred Hall, Staff Associate, Eisenhower Regional Consortium at Appalachian Educational Laboratory

Connie Hames, Geology Teacher, Stafford County Public Schools, Virginia

Jayne Hart, Professor of Biology, George Mason University

Michelle Kipke, Director, Forum on Adolescence, Institute of Medicine

John Layman, Professor Emeritus of Physics, University of Maryland

Thomas Liao, Professor and Chair, Department of Technology and Society, State University of New York at Stony Brook

Ian MacGregor, Director, Division of Earth Sciences, National Science Foundation

Ed Mathews, Physical Science Teacher, Fairfax County Public Schools, Virginia

Ted Maxwell, Geomorphologist, National Air and Space Museum, Smithsonian Institution

Tom O'Haver, Professor of Chemistry/Science Education, University of Maryland

Robert Ridky, Professor of Geology, University of Maryland

Mary Alice Robinson, Science Teacher, Stafford County Public Schools, Virginia

Bob Ryan, Chief Meteorologist, WRC Channel 4, Washington, D.C.

Michael John Tinnesand, Head, K-12 Science, American Chemical Society

Grant Woodwell, Professor of Geology, Mary Washington College

Thomas Wright, Geologist, National Museum of Natural History, Smithsonian Institution; U.S. Geological Survey (emeritus)

Acknowledgments

The National Science Resources Center gratefully acknowledges
the following individuals and school systems for their assistance
with the national field-testing of *Catastrophic Events:*

East Bay Educational Collaborative, Rhode Island

Site Coordinator
Ronald D. DeFronzo, Science Specialist
East Bay Educational Collaborative
Director, Kits in Teaching Elementary Science,
Portsmouth

Michael J. Brennan, Teacher
Portsmouth Middle School, Portsmouth

Mary J. Hayes, Teacher
Thompson Middle School, Newport

Donna Stouber, Teacher
Kickemuit Middle School, Warren

**School District of Greenville County
Greenville, South Carolina**

Site Coordinator
Toni Enloe, Teaching and Learning Division

Elayne R. Finkelstein, Teacher
League Academy

Robbie L. Higdon, Teacher
League Academy

Mary Helen Maxwell, Teacher
League Academy

**Minneapolis Public Schools
Minneapolis, Minnesota**

Site Coordinator
James Bickel, Teacher and Instructional
Services

Ann Ginis, Teacher
Benjamin Banneker Community School

Michael Madden, Teacher
Ann Sullivan Communication Center

Holly C. Thompson, Teacher
Franklin Middle School

**Montgomery County Public Schools
Montgomery County, Maryland**

Site Coordinator
Patricia A. Hagan
Middle School Science Specialist

Theresa Manley Sykes,
Science Resource Teacher
White Oak Middle School

**School District of Philadelphia
Philadelphia, Pennsylvania**

Site Coordinator
Allen Ruby, Research/Curriculum Specialist,
Talent Development Schools,
Center for Social Organization of Schools,
Johns Hopkins University

Deborah Bambino, Teacher
Central East Middle Annex

Jacqueline Dubin, Teacher
Jay Cooke Middle School

Donald L. Rissover, Teacher
Beeber Middle School

Redwood City School District
Redwood City, California

Site Coordinator
Dorothy Patzia, Science Resource Teacher
Bay Area Schools for Excellence in Education
(BASEE)

Anne Renoir, Teacher
Garfield Charter Middle School, Menlo Park

Sandra Robins, Teacher
Hoover Math and Tech Magnet, Redwood City

Bobbie Stumbaugh, Teacher
Selby Lane School, Atherton

Stafford County Public Schools
Stafford County, Virginia

Site Coordinator
Barry Mathson, Science Coordinator

Jan Pierson, Teacher
Gayle Middle School

Winston Ward, Principal
Gayle Middle School

Michael Wondree, Assistant Principal
Gayle Middle School

The NSRC thanks the following individuals for their assistance
during the development of *Catastrophic Events*:

Jody Hayob, Geology Professor
Mary Washington College
Fredericksburg, Virginia

Maureen Kerr
Educational Services Manager
National Air and Space Museum
Educational Services
Smithsonian Institution, Washington, D.C.

Fred Klein, Seismologist
U.S. Geological Survey
Menlo Park, California

James F. Luhr, Curator
Global Volcanism Project
National Museum of Natural History
Smithsonian Institution, Washington, D.C.

Steven Mabry, Electronics Engineer
Technology Management Group, Inc.
Dahlgren, Virginia

Amanda May, Teacher
Mountain View Elementary
Haymarket, Virginia

Charles J. Pitts, Electrical Engineer
Science Application International Corporation
McLean, Virginia

Dennis Schatz, Associate Director
Pacific Science Center, Seattle, Washington

Tom Simkin, Curator
National Museum of Natural History
Smithsonian Institution, Washington, D.C.

Rose Steinet, Photo Librarian
Center for Earth and Planetary Studies
National Air and Space Museum
Smithsonian Institution, Washington, D.C.

Terry Teays, Manager of Education Group
Origins Education Forum Scientist
Space Telescope Science Institute
Baltimore, Maryland

Penny Sullivan
American Rescue Dog Association
New York, New York

Tim Watts, Teacher
Chemistry and Marine Science
Courtland High School
Spotsylvania County Public Schools
Spotsylvania, Virginia

The NSRC appreciates the contribution of its
STC/MS project evaluation consultants—

Program Evaluation Research Group (PERG), Lesley College

Sabra Lee
Researcher, PERG

George Hein
Director (retired), PERG

Center for the Study of Testing, Evaluation,
and Education Policy (CSTEEP), Boston College

Joseph Pedulla
Director, CSTEEP

The NSRC acknowledges the contributions of
the Smithsonian Center for Education and Museum Studies—

The content of the inquiry on climate signals in Lesson 11 and the reading selection on
prehistoric climate change were adapted from educational materials they developed.

Preface

Community leaders and state and local school officials across the country are recognizing the need to implement science education programs consistent with the National Science Education Standards to attain the important national goal of scientific literacy for all students in the 21st century. The Standards present a bold vision of science education. They identify what students at various levels should know and be able to do. They also emphasize the importance of transforming the science curriculum to enable students to engage actively in scientific inquiry as a way to develop conceptual understanding as well as problem-solving skills.

The development of effective standards-based, inquiry-centered curriculum materials is a key step in achieving scientific literacy. The National Science Resources Center (NSRC) has responded to this challenge through Science and Technology Concepts-Secondary™. Prior to the development of these materials, there were very few science curriculum resources for secondary students that embodied scientific inquiry and hands-on learning. With the publication of STC-Secondary™, schools will have a rich set of curriculum resources to fill this need.

Since its founding in 1985, the NSRC has made many significant contributions to the goal of achieving scientific literacy for all students. In addition to developing Science and Technology Concepts-Elementary™—an inquiry-centered science curriculum for grades K through 6—the NSRC has been active in disseminating information on science teaching resources, preparing school district leaders to spearhead science education reform, and providing technical assistance to school districts. These programs have had a significant impact on science education throughout the country. The transformation of science education is a challenging task that will continue to require the kind of strategic thinking and insistence on excellence that the NSRC has demonstrated in all of its curriculum development and outreach programs. The Smithsonian Institution, our sponsoring organization, takes great pride in the publication of this exciting new science program for secondary students.

Letter to the Students

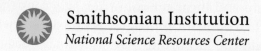

Smithsonian Institution
National Science Resources Center

Dear Student,

The National Science Resources Center's (NSRC) mission is to improve the learning and teaching of science for K-12 students. As an organization of the Smithsonian Institution, the NSRC is dedicated to the establishment of effective science programs for all students. To contribute to that goal, the NSRC has developed and published two comprehensive, research-based science curriculum programs: Science and Technology Concepts-Elementary™ and Science and Technology Concepts-Secondary™.

By using the STC-Secondary™ curriculum materials, we know that you will build an understanding of important concepts in life, earth, and physical sciences; learn critical-thinking skills; and develop positive attitudes toward science and technology. The National Science Education Standards state that all secondary students "...should be provided opportunities to engage in full and partial inquiries.... With an appropriate curriculum and adequate instruction, ... students can develop the skills of investigation and the understanding that scientific inquiry is guided by knowledge, observations, ideas, and questions."

STC-Secondary also addresses the national technology standards published by the International Technology Education Association. Informed by research and guided by standards, the design of the STC-Secondary units addresses four critical goals:

• Use of effective student and teacher assessment strategies to improve learning and teaching

• Integration of literacy into the learning of science by giving students the lens of language to focus and clarify their thinking and activities

• Enhanced learning using new technologies to help students visualize processes and relationships that are normally invisible or difficult to understand

• Incorporation of strategies to actively engage parents to support the learning process

We hope that by using the STC-Secondary curriculum you will expand your interest, curiosity, and understanding about the world around you. We welcome comments from students and teachers about their experiences with the STC-Secondary program materials.

Thomas Emrick
Executive Director
National Science Resources Center

Navigating an STC–Secondary™ Student Guide

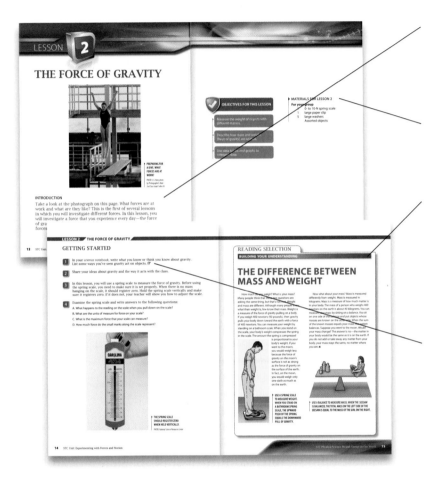

INTRODUCTION
This short paragraph helps get you interested about the upcoming inquiries.

MATERIALS
This helps you get organized and prepare for your inquiries.

READING SELECTION:
BUILDING YOUR UNDERSTANDING
These reading selections are part of the lesson, and give you information about the topic or concept you are exploring.

NOTEBOOK ICON
During the course of an inquiry, you'll record data in different ways. This icon lets you know to record in your science notebook. Student sheets are called out when you're to write there. You may go back and forth between your notebook and a student sheet. Watch carefully for the icon throughout the procedure.

SAFETY TIPS
Safety in the science classroom is very important. Tips throughout the student guide will help you to practice safe techniques while conducting investigations. It is very important to read and follow all safety tips.

SAFETY TIP

PROCEDURE
This tells you what to do. Sometimes the steps are very specific, and sometimes they guide you to come up with your own investigation and ways to record data.

REFLECTING ON WHAT YOU'VE DONE
These questions help you think about what you've learned during the lesson's inquiries, apply them to different situations, and generate new questions. Often you'll discuss your ideas with the class.

READING SELECTION: EXTENDING YOUR KNOWLEDGE
These reading selections come after the lesson, and show new ways that the topic or concept you learned about during the lesson can be applied, often in real-world situations.

GLOSSARY
Here you can find scientific terms defined.

INDEX
Locate specific information within the student guide using the index.

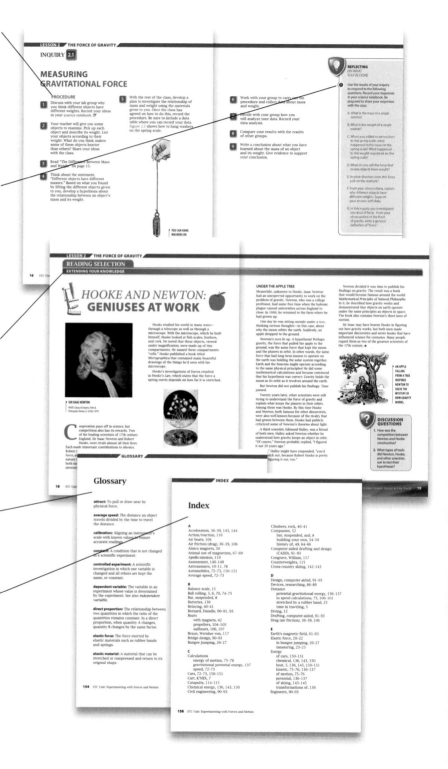

Contents

CONTENTS

PART 3:
CLIMATE AND CLIMATE CHANGES

OUR IDEAS ABOUT WEATHER AND CLIMATE

▶ **REBUILDING LOUISIANA AFTER HURRICANE KATRINA**

PHOTO: NOAA/Collection of Wayne and Nancy Weikel, FEMA Fisheries Coordinators

INTRODUCTION

The earth is a dynamic complex of four systems: its waters, air, rock, and living creatures. We know that change in one earth system affects the others— but how? Geologists, meteorologists, oceanographers, ecologists, climatologists, paleontologists, and even social scientists study these complex relationships.

In this unit, you will investigate how oceans and atmosphere interact as they contribute to earth's weather and climates. You will seek the causes of storms, winds, and catastrophic events like hurricanes and tornadoes. You will study the oceans and how they exchange heat with the atmosphere. Finally, you will examine earth's climates and evidence for projected climate change.

To begin, you will use a map and globe to mark places where you think major weather and climate-related phenomena occur. What do you already know? Let's find out.

OBJECTIVES FOR THIS LESSON

Independently and collaboratively record ideas and questions about earth's weather and climate.

Compare the features of a globe with those of a map.

Record where you think weather and climate events occur on the earth.

▶ **MATERIALS FOR LESSON 1**

For you

1   copy of Student Sheet 1.2a: Pre-Unit Assessment: Where on Earth?

For your group

1   large sheet of paper or newsprint
1   set of colored markers
1   inflatable globe
1   set of multicolored dots
1   Weather and Climate World Map

GETTING STARTED

1 Read the Introduction and the Objectives for this lesson.

2 Discuss with the class what you consider to be a naturally occurring event in the earth's atmosphere.

3 In your science notebook, create a concept map using the terms "weather" and "climate." Record what you know about weather and climate and their relationship by using as many descriptive terms as possible. 🖉

▶ A WEATHER BALLOON IS RELEASED AT CAPE CANAVERAL WEATHER STATION IN FLORIDA. ITS JOB IS TO MONITOR WIND SPEED AND DIRECTION IN THE ATMOSPHERE PRIOR TO THE LAUNCH OF A SPACE SHUTTLE.

PHOTO: NASA

CREATING A GROUP CONCEPT MAP

PROCEDURE

1 Get together with your group. Your group will work together throughout the unit. It will be important to work collaboratively with other group members to obtain the best experimental results and promote your own learning.

2 Have someone from your group pick up a large sheet of paper and a set of colored markers. Share with your group the concept map on weather and climate you recorded in your notebook. Working together, create a group concept map on the large sheet of paper. Remember to date and label the sheet. Put your names and class period on it.

3 Be prepared to share your group's concept map with the class.

READING SELECTION

BUILDING YOUR UNDERSTANDING

IMAGES OF THE EARTH

Clouds usually cover about 50 percent of the earth's surface. Photographs, such as the one below, show how the earth looks from space. Look closely at the image. You can see the earth and some of its seven continents. Vast oceans surround the continents. Clouds swirl above them.

During the lessons in this unit, you will use a globe and a map of the world to make and record some of your observations of the earth and its atmospheric and oceanic processes and climate zones. A globe is a spherical model of the earth in three dimensions. Your map shows the earth, or parts of it, in two dimensions on a flat surface.

The globe you will use during this unit is an inflatable globe that shows the earth as it looks from space. An artist created the globe by using hundreds of satellite images and photographs taken from space, as well as geophysical maps that show mountain ranges and deserts. The artist used images from the Northern Hemisphere in autumn and images of the Southern Hemisphere in the spring. Images taken at those times of year contain less cloud cover and make the continents more visible. ■

▶ **WHICH CONTINENTS CAN YOU RECOGNIZE IN THIS IMAGE OF EARTH?**

PHOTO: NASA Goddard Space Flight Center Scientific Visualization Studio/Blue Marble data courtesy of Reto Stockli (NASA/GSFC)

USING A GLOBE AND A WORLD MAP

PROCEDURE

1 Read "Images of the Earth" and think about what the globe for this inquiry will look like.

2 Record your ideas about weather and climate events on earth on Student Sheet 1.2a: Pre-Unit Assessment: Where on Earth? Complete Table 1 on your own. Return the completed sheet to your teacher.

3 Your teacher will give your group an inflatable globe. Look at it carefully. Discuss your general observations with your group. How is the globe different from the image of the earth on page 6?

4 Answer the following questions in complete sentences in your science notebook. Be ready to share your ideas with the class. 📝

A. Is there any evidence on the globe that the earth's surfaces are active (or changing)? If so, describe it.

B. How might you use this globe to learn more about the concepts you discussed in Inquiry 1.1?

C. How might photographs of Earth taken from space help scientists predict or monitor atmospheric or oceanic processes?

5 When your group receives the Weather and Climate World Map, locate your continent, then your country, and your particular location in the country. Identify the other continents and as many countries as you can. How is your map like the globe? How is it different? How is the scale on the globe different from the scale on the map?

Inquiry 1.2 continued

6 Return the globe to your teacher. Get one set of multicolored dots. As a group, use the dots and the following key to record on your map where you think major atmospheric events and oceanic processes occur most frequently.

KEY

• Tornado—orange

• Hurricane—green

• Flood or drought—red

• Ice melt—yellow

7 Be prepared to share your group's map with the class.

8 Clean up your materials. Your group's map should be labeled with your group's names and class period.

REFLECTING
ON WHAT
YOU'VE DONE

1 Answer the following questions in your science notebook:

A. Are any of the weather and climate events that you listed in your notebook or recorded on your world map related to each other in any way? If so, how?

B. Do any of these events change the way the earth looks over time? If so, which ones? Why do you think this happens?

C. How do you think weather and climate affect people?

D. How might scientists predict the weather and climate?

2 Record in your science notebook any questions you may have about weather and climate. Share your questions with the class. As you complete each lesson, try to find answers to these questions.

How Scientists Study the Earth

▶ A SCIENTIST SETS UP A STATION THAT WILL MONITOR
CORAL CALCIFICATION OFF THE COAST OF FLORIDA.

PHOTO: U.S. Geological Survey/photo by Ilsa B. Kuffner

Throughout this unit, you'll read about what scientists know of the earth's weather, climate, and oceans. But how do they know? How do scientists learn about the earth and the things that cause it to change over time?

Scientists break down each earth system (the atmosphere, the hydrosphere, the geosphere, and the biosphere) into many different fields of study. In this unit, you'll read about the work of atmospheric scientists, oceanographers, climatologists, and paleobiologists.

▶ TWO SCIENTISTS OBSERVE ARCTIC SEA ICE WHILE ON AN EXPEDITION TO MAP THE ARCTIC SEAFLOOR.

PHOTO: U.S. Geological Survey/photo by Jessica K. Robertson

▶ THIS METEOROLOGIST IS SETTING UP A PORTABLE ALL HAZARDS METEOROLOGICAL RESPONSE SYSTEM (AMRS). THESE SYSTEMS ARE USED DURING EMERGENCIES, SUCH AS WILDFIRES, TO PROVIDE ACCESS TO WEATHER RADAR, COMPUTER FORECAST MODELS, AND SATELLITE IMAGES.

PHOTO: NOAA

Atmospheric scientists study the chemical makeup of the atmosphere and its dynamics (how it moves and why). They also study how the atmosphere interacts with the rest of the planet. If the carbon dioxide concentration in the atmosphere increases, for instance, how might it affect the oceans?

Meteorologists are one type of atmospheric scientist. They monitor, investigate, and forecast the weather, which is very important if you're trying to stay out of the way of a hurricane or tornado, are growing crops, or are just trying to figure out what to wear to school.

Another kind of atmospheric scientist is a climatologist, or climate scientist. Climatologists study the weather in large areas over long spans of time—anything from years to billions of years—and try to understand the causes of differences in climates around the globe, as well as climate change.

Oceanographers study the ocean, and you can see their field's origin in its name: "-graphy" means "writing," or charting. Early oceanographers mapped the ocean, not just so they could travel safely, but so they could find out where the oceans ended, and what, if anything, was at the end of them.

Modern oceanographers study various aspects of the ocean. Marine physicists study motion in the ocean: currents and waves, tides, and conveyers, which are large belts of water carrying heat around the globe. Marine geologists study the fossil evidence of the seafloor and undersea plate tectonics, including the evolution of deepwater trenches and oceanic mountain ranges. Marine chemists study the chemistry of the oceans: seawater composition at various depths, and how surface waters heat and evaporate. And marine biologists study the life of the ocean.

Because 71 percent of the globe is covered by oceans, the oceanographers' work is crucial to understanding the earth and its environments. There's a great deal of work to be done, too. Most of the undersea world remains a mystery to us.

HOW THEY DO THEIR WORK

Atmospheric scientists and oceanographers collect data, try to understand what it tells them about the skies and oceans, and use that knowledge to build rich, dynamic models of the earth. In this unit, you'll read about climatologists' use of very old fossil evidence and very new satellite data. You'll also read about meteorologists' centuries-old use of mercury for studying air pressure, and their more modern instruments. Oceanographers, too, have an ingenious collection of tools: they've used everything from messages in bottles to sensitive tests of ocean-water temperature, gas content, and salt concentration.

KNOWING WHICH DATA TO COLLECT, AND BUILDING THE TOOLS TO GET IT

Much of science is about figuring out exactly what your question is and what you need to know in order to answer it. Do you need to know the speed of the wind 200 feet over the prairie? The way that solar radiation scatters when it hits certain molecules in the air? The path of an ocean current? And what does "the path of an ocean current" mean—where does the width of a current start and end? Do you need measurements, and if so, what kind?

Once scientists define precisely what they need to know, they can design and build instruments to collect the data. To do that they use chemistry, biology, physics, and engineering. For instance, a meteorologist who's trying to understand how winds change in an area just before a tornado comes through might want a sensitive wind detector that swivels very easily, picking up subtle shifts in the wind but not spinning when birds fly by. He might need to figure out the amount of friction in the swivel joint, the breezes' ability to perform work on the instrument's moving parts—and how strong a breeze a bird generates. He would also need to find a way for the instrument to record its measurements.

READING SELECTION
EXTENDING YOUR KNOWLEDGE

▶ A RESEARCHER DRILLS TO OBTAIN AN ICE CORE IN THE MCMURDO DRY VALLEYS IN ANTARCTICA. THE ICE CORE WILL PROVIDE CLUES INTO THE HISTORY OF THE AREA'S CLIMATE.

PHOTO: Jennifer Heldmann/National Science Foundation

Once they've got their instruments, scientists must figure out how to get them to the right parts of the earth. Sometimes that means putting the instrument on a satellite and flying it into orbit; sometimes it means putting it into a bathyscape for a journey deep undersea. And sometimes it just means clipping the instrument to a pole in the backyard.

If you become a scientist and do your own research on pressing scientific questions, you may find that your work involves not just using tools, but inventing new ones.

WHEN THE DATA IS COLLECTED

Computers can be programmed to turn data into easy-to-read forms and reports. Ultimately, though, the scientists must decide what the data mean. This isn't easy! Whatever information they get is only a tiny slice of what there is to know about the phenomenon they're studying. Carbon dioxide–monitoring satellites, for instance, don't scan the entire Earth at every moment to find out where carbon dioxide comes from and goes. They scan tiny, carefully chosen slices of the atmosphere. Scientists must try to connect their dots of data to make plausible models of what goes on in the oceans and atmosphere.

These models help them advise government officials on building codes, emergency action plans, climate forecasts, and more. In some cities, for instance, governments must now tell builders how climate change might affect their new buildings. Other cities wait to hear whether they'll need to move entire neighborhoods out of a rising ocean's way. In other words, the science you'll read about here isn't just amazing—it's urgent. ■

DISCUSSION QUESTIONS

1. Why are there so many kinds of atmospheric and oceanographic scientists?

2. How is the work of a scientist like that of an explorer, an inventor, an engineer, and a writer?

VIEWS FROM SPACE

▶ **AN AERIAL VIEW OF LOS ANGELES, TAKEN FROM A BALLOON IN 1902**

PHOTO: Library of Congress, Prints & Photographs Division, PAN US GEOG — California no. 269

The urge to view ourselves, our cities, and our planet from afar is irresistible. Every time technology improves, we try to view our world from a new perspective, and hope to learn something surprising. In 1860, James W. Black, a Boston photographer, took what is believed to be the first successful image of the earth from the sky. He was in a tethered balloon 370 meters (1214 feet) high when he shot the picture.

Early aerial photographers not only took to balloons to get their cityscapes, they also fastened cameras with shutters on timers to kites, rockets, and even carrier pigeons.

When the airplane was invented in the early 1900s, more controlled aerial photography became possible. No longer just a curiosity for surveying or postcard views, aerial photography could be used for assessing damage after an earthquake, tracking the movement of glaciers, observing erosion along a coastline, mapping forests, and tracking fires. It was also used for military purposes, surveying landscapes without sending scouts into potentially dangerous terrain.

Today, aerial photography is vastly more sophisticated. For example, a process called LIDAR (Light Detection And Ranging), bounces laser pulses, rather than sound, off objects in the landscape and atmosphere—including small objects such as raindrops or rocks. The light's return time allows the LIDAR apparatus to calculate how far away objects are, and use that data to generate a terrain map. This technique is useful in the study of atmospheric sciences, seismology, geology, forestry, archaeology, meteorology, and landforms and topography.

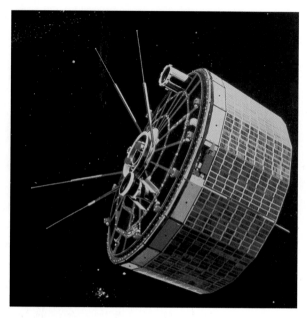

▶ **THE UNITED STATES LAUNCHED ITS FIRST WEATHER SATELLITE, TIROS 1, IN 1960.**

PHOTO: NASA

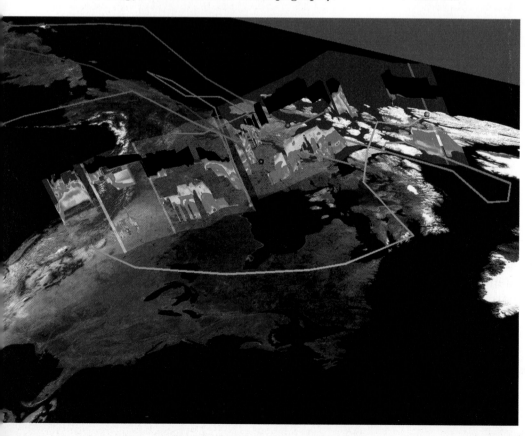

▶ **THIS GRAPHIC, BASED ON DATA COLLECTED BY LIDAR EQUIPMENT, SHOWS OZONE LEVELS IN THE ATMOSPHERE. BLUE AREAS HAVE LOW OZONE LEVELS, WHILE YELLOW TO RED AREAS HAVE HIGH LEVELS. BLACK AREAS REPRESENT THE STRATOSPHERE, WHERE OZONE LEVELS ARE VERY HIGH.**

PHOTO: Image courtesy of Kurt Severance, NASA Langley Research Center

▶ IMAGE OF EARTH FROM SPACE AS
CAPTURED BY THE APOLLO 8 CREW

PHOTO: NASA

PHOTOGRAPHS FROM SPACE

In 1968, Apollo 8 astronaut William Anders took the first and arguably the most famous picture of Earth seen from space, dubbed "Earthrise." While beautiful—so beautiful that it's credited with starting the modern environmental movement—it was not taken just to capture physical beauty. Astronomers began taking photos from space because they wanted to understand our atmosphere.

In the early 1960s, the National Aeronautics and Space Administration (NASA) launched the first satellite designed to see what Earth looked like from space. TIROS 1 (or Television Infrared Observing Satellite) was a weather satellite: an instrument that orbits Earth, taking photographs and collecting measurements. It took the first pictures of Earth from 640 kilometers (398 miles) above its surface. These photographs confirmed what meteorologists had suspected: clouds in the atmosphere have patterns that look like swirls, bands, or clusters.

This was important information. Before the invention of weather satellites, scientists could not detect severe storms until the storms moved dangerously close to populated areas. People living in the path of a hurricane, for example, got almost no warning. Understanding cloud patterns and the motion of storms across the face of the planet helped meteorologists predict weather days ahead with far greater accuracy. Today, satellites can predict, locate, and track hurricanes and tornadoes while they're still far from population centers. Scientists transmit this information to public officials, who can warn people of the storm and tell them to evacuate or take cover if necessary.

READING SELECTION

▶ NASA'S TERRA SATELLITE IS COLLECTING DATA TO HELP SCIENTISTS BETTER UNDERSTAND CLIMATE AND ENVIRONMENTAL CHANGE. ITS MODIS (MODERATE RESOLUTION IMAGING SPECTRORADIOMETER) INSTRUMENT CAPTURED THIS IMAGE OF HURRICANE RITA AS IT ENTERED THE MISSISSIPPI VALLEY FROM THE GULF OF MEXICO.

PHOTO: Jeff Schmaltz/NASA Visible Earth

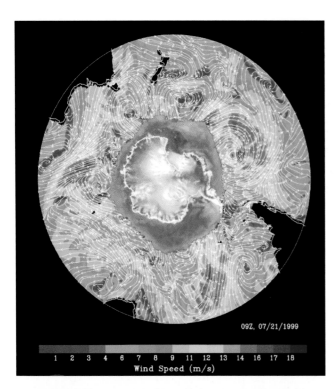

▶ THE SEAWINDS INSTRUMENT ON NASA'S QUICKSCAT SATELLITE MONITORS SEA ICE AND OCEAN SURFACE WIND TO BETTER UNDERSTAND HOW THEY INTERACT TO AFFECT CLIMATE. THIS IMAGE SHOWS ANTARCTIC ICE AND WIND. THE GRAY AREA AROUND ANTARCTICA REPRESENTS SEA ICE. THE WHITE LINES SHOW SURFACE WIND DIRECTION AND THE COLORS REPRESENT WIND SPEED.

PHOTO: NASA Jet Propulsion Laboratory/NASA Visible Earth

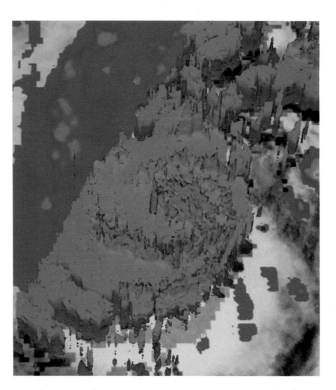

▶ NASA'S TRRM (TROPICAL RAINFALL MEASURING MISSION) SATELLITE SENDS BACK INFORMATION ON THE AMOUNT OF RAINFALL OVER 24 HOURS FOR A GIVEN AREA. THESE DATA HELP METEOROLOGISTS PREDICT TROPICAL STORMS AND HURRICANES MORE ACCURATELY. THIS IMAGE SHOWS RAINFALL RATES DURING HURRICANE IRENE. RED AREAS INDICATE INTENSE RAINFALL. THE RED TOWERS ARE RAIN COLUMNS RISING ABOVE THE OCEAN'S SURFACE. THESE STORMS-WITHIN-A-STORM ARE OFTEN A SIGN THAT A HURRICANE IS INTENSIFYING.

PHOTO: Images produced by Hal Pierce. Caption by Steve Lang and Hal Pierce/NASA Visible Earth

SATELLITES IN ORBIT

How does a weather satellite get into orbit? A rocket launches it to an altitude high enough so that, if it moves at an appropriate speed, Earth's gravitational force keeps it in orbit. If it moves too slowly, gravity will pull it back to Earth. If the satellite moves too fast, it will escape Earth's gravitational pull and fly off on a tangent into space.

▶ AN ARTIST'S VERSION OF THE SOLAR AND HELIOSPHERIC OBSERVATORY (SOHO) SATELLITE LAUNCHED IN 1995. SOHO HAS 12 INSTRUMENTS ABOARD COLLECTING INFORMATION ABOUT THE STRUCTURE AND DYNAMICS OF THE SOLAR INTERIOR, THE SOLAR CORONA, AND SOLAR WIND.

PHOTO: Courtesy of SOHO consortium. SOHO is a project of international cooperation between ESA and NASA.

▶ ENVIRONMENTAL SATELLITES ALSO MONITOR A VARIETY OF OTHER INFORMATION, SUCH AS THE STATUS OF VOLCANOES AND FIRES, OCEAN TEMPERATURE, AND DEFORESTATION. THIS SATELLITE IMAGE, CAPTURED BY THE MODIS INSTRUMENT FROM NASA'S AQUA SATELLITE, SHOWS ACTIVELY BURNING FIRES IN MEXICO AND CENTRAL AMERICA IN RED. A THICK LAYER OF SMOKE IS ALSO EVIDENT IN THE IMAGE.

PHOTO: Jeff Schmaltz/NASA Visible Earth

Scientists primarily use two kinds of satellites for viewing clouds: the Geostationary Orbiting Environmental Satellite, often called GOES, and the Polar Orbiting Environmental Satellite, called POES. Once in space, these satellites do not change their orbits around Earth unless they receive a command to do so.

A geostationary satellite is located about 35,000 kilometers (about 22,000 miles) above the equator. It orbits Earth's axis in exactly the same time and the same direction as Earth rotates on its axis, traveling with the planet's rotation. Seen from Earth, the satellite is always in the same position, always observing the same area.

Polar orbiting satellites are positioned about 850 kilometers (about 530 miles) above Earth. They have pole-to-pole orbits, moving north and south. Because Earth rotates but the north-south orbit of a polar satellite is static, this satellite can image section after section of Earth as the planet rotates beneath it.

IMAGES OF EARTH

Equipped with a wide array of sensors, satellites can measure cloud cover, ice cover, temperature, air pressure, precipitation, and the chemical composition of the atmosphere, then transmit data measurements to receiving stations on Earth. The data is analyzed and sent to forecasters and other scientists; some data is sent in real time, allowing scientists to make on-the-spot assessments of weather and other situations.

Satellites are used not just for scientific research on Earth, but also other planets. With the aid of satellites, scientists can gather information about Earth and our solar system that was once thought beyond human reach. Satellites have changed how we study Earth, what we know about the solar system, and what we can imagine learning in the future. ■

(A) GOES

(B) POES

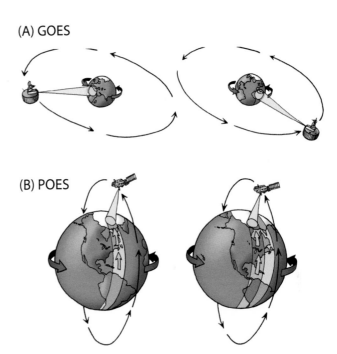

▶ (A) GOES' ORBIT IS IN SYNC WITH EARTH'S ROTATION. IT CONSTANTLY VIEWS THE SAME AREA OF EARTH AND REMAINS AT THE SAME POINT OVER THE EQUATOR. (B) POES SCANS EARTH FROM NORTH TO SOUTH. AS EARTH ROTATES ON ITS AXIS, THE SATELLITE IS ABLE TO SCAN AN AREA FARTHER TO THE WEST WITH EACH PASS.

 DISCUSSION QUESTIONS

1. Why do you think the Earthrise photo had such a big effect on people who saw it in 1968?

2. What sorts of information can we gather about Earth that we could not have gathered before the 20th century?

INTRODUCING STORMS

THE POWERFUL WINDS OF HURRICANE ANDREW DROVE THIS PIECE OF PLYWOOD THROUGH THE TRUNK OF A ROYAL PALM TREE.

PHOTO: National Oceanic and Atmospheric Administration (NOAA)/Department of Commerce

INTRODUCTION

Hurricanes and other rotating storms are strong forces of nature. Both hurricanes and tornadoes have winds powerful enough to lift objects and turn them into dangerous, high-speed missiles, as shown in the photograph.

In Lessons 5, 6, and 7 you will study the origins and behavior of these catastrophic storms, as well as their effects on the areas they strike and the people who live there. In this inquiry, you will study something fundamental to both hurricanes and tornadoes: the formation and behavior of vortices, those tight, spinning, tornado-like forms. ("Vortices" is the plural of "vortex.") You will use two bottles—one empty and one filled with water—to model how air moves in hurricanes, tornadoes, and other rotating storms. You'll discuss how the water in your bottle is like the air in a rotating storm as you model the circulation in a storm. You'll also take a step back and read about thunderstorms, whose powerful updrafts can be a nursery for vortices of air and full-fledged tornadoes.

OBJECTIVES FOR THIS LESSON

▶ View satellite images of clouds and identify movement patterns within the clouds.

▶ Model the movement of air in a tornado or hurricane.

▶ Create a working definition for the word "vortex."

▶ Read to learn more about thunderstorms, tornadoes, and hurricanes.

▶ **MATERIALS FOR LESSON 2**

For your group

1	plastic box with lid
	4 paper towels
	1 Weather and Climate World Map
	1 set of multicolored dots
	1 group concept map (from Lesson 1)
1	set of assorted colored pencils, crayons, or markers
1	vortex model

GETTING STARTED

1 Think back to Lesson 1. Where did your group think most tornadoes and hurricanes occurred? Share with the class the locations your group chose and explain why.

2 Examine Figures 2.1 and 2.2. Then try to answer the questions below with your group and class.

A. Think about the reading selection "Views From Space" in Lesson 1. How do you think the images in Figures 2.1 and 2.2 were made?

B. How are the two images alike?

C. How are they different?

D. What patterns do you notice in the shapes of the clouds? Why do you think these patterns form?

▶ **HURRICANE LINDA AS IT APPROACHED BAJA CALIFORNIA, MEXICO.**
FIGURE **2.1**

PHOTO: NASA Goddard Space Flight Center/NOAA

▶ **A COMPUTER-ENHANCED VIEW OF HURRICANE LINDA AS IT APPROACHED BAJA CALIFORNIA, MEXICO.**
FIGURE **2.2**

PHOTO: NASA Goddard Space Flight Center/NOAA

MODELING A VORTEX

PROCEDURE

1 Watch as your teacher holds up the class vortex model. What happens to the water? Discuss your observations with the class. How can you explain what you see?

2 Watch as student volunteers practice getting the water in the top bottle to move quickly into the bottom bottle.

3 Now your group will try this. Review Procedure Steps 4 through 7 with your teacher. Then pick up your group's materials.

4 Just as your teacher did, turn the model so that the water is in the top bottle. Hold the bottle very still. What happens? (Work above your plastic box in case of spills.)

5 Try to get the water in the top bottle to flow into the bottom bottle. What must you do to make this happen? Investigate with your group. Do this several times.

READING SELECTION

BUILDING YOUR UNDERSTANDING

TORNADO WATCH OR WARNING?

What is a "tornado watch"? What is a "tornado warning"? Both these terms alert the public that a threatening weather system is approaching, but they signify different levels of danger.

A tornado watch means that tornadoes are possible, given the atmospheric conditions and storm clouds in the area. Thunderstorms with high winds and heavy rain, for instance, may produce a tornado. Radar patterns that show the possibility of cloud rotation, such as a hooked shape to a line of storms, are also signs to watch. Residents under a tornado watch should be alert to the possibility of a tornado warning coming at any moment.

A tornado warning means that a tornado has actually been spotted or detected by radar. People in the area should get indoors and stay away from windows and outside walls. They must take shelter immediately by going to their basement or the lowest part of their home. If they have no basement, they can go to a bathroom or a closet, particularly a closet under a staircase.

Preferably, people should choose an area well inside the house, away from outside walls; the main danger in a tornado comes from flying debris such as tree branches, which high winds can fling through walls with surprising speed and ease. A bathtub can be a safe place to wait out a tornado. Once in your safe spot, protect your head and neck by covering up with cushioning and blankets. And don't forget a phone! Should a tornado strike your building, you'll want to be able to let people know where you are.

▶ HOW DOES WEATHER INFORMATION GET TO THE PUBLIC? THIS IS A SIMPLIFIED VERSION OF HOW SCIENTISTS COLLECT, PROCESS, AND DELIVER SEVERE-WEATHER DATA TO THE PUBLIC.

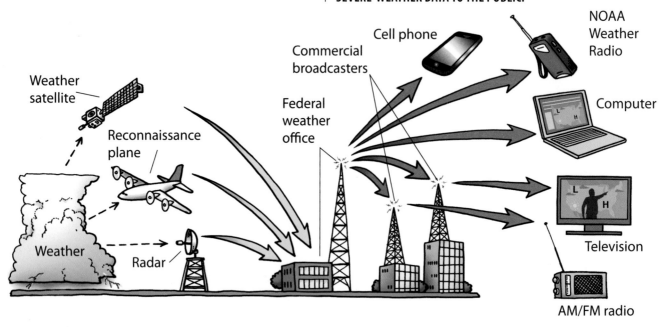

Weather satellite · Reconnaissance plane · Weather · Radar · Federal weather office · Commercial broadcasters · Cell phone · NOAA Weather Radio · Computer · Television · AM/FM radio

How do you know when a watch or warning for tornadoes has been issued? Local television stations and the National Weather Service broadcast tornado watches and warnings on the Internet, television, and the radio. In some areas, you can have tornado watch alerts sent to your phone or email address. You can also listen for tornado news on a weather radio, which is a special radio that broadcasts warnings and information for all types of hazards: weather (e.g., floods, hurricanes, tornadoes), natural disasters (e.g., earthquakes, forest fires, and volcanic activity), technological catastrophes (e.g., chemical releases, oil spills, nuclear power plant emergencies, etc.), and other national emergencies.

In many areas where tornadoes are common, warning sirens let people know when a tornado has been spotted. These sirens usually sound the all-clear, too, after a tornado has passed. To respond to a warning, follow the safety instructions given by the system.

Finally, use your own eyes and watch the skies when conditions are ripe for tornadoes. A telltale greenish cast to the sky precedes many tornadoes, and so do extremely hot, humid, still conditions followed by a sudden violent thunderstorm with cold gusts pushing through. If you watch the sky as tornado-generating storms gather and pass through, you may see rotation in the clouds, or the formation of a "wall cloud," which drops down like a fat mushroom stem from the cloud above. Don't stand outside watching the storm too long, though; if a siren or other alert sounds, be ready to get inside and take cover. ■

6 Observe what happens to the glitter and beads when the water moves in a spiral. Discuss the following questions with your group as you work:

A. Where are most of the glitter and beads?

B. Where is the movement of the glitter and beads the fastest? Where is it the slowest?

C. How is the vortex model like a real tornado? How is it like a hurricane?

7 Clean up. Your teacher will tell you what to do with the bottles.

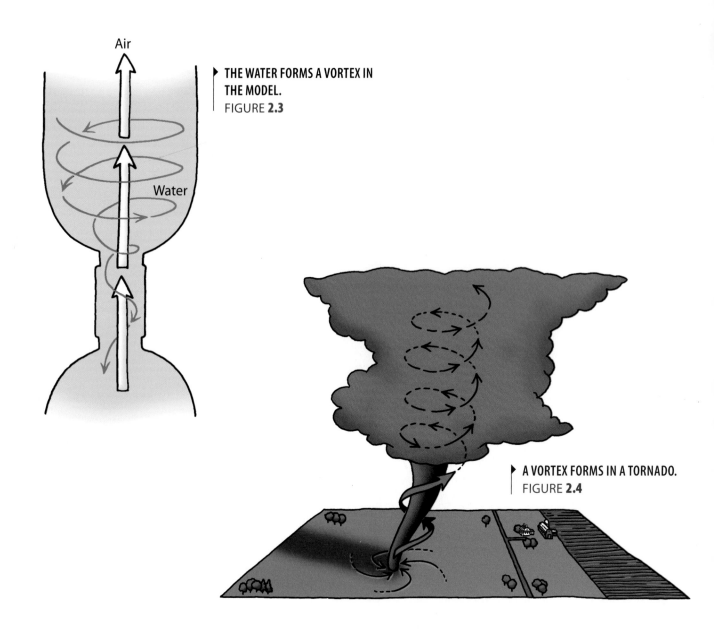

Air

Water

▶ **THE WATER FORMS A VORTEX IN THE MODEL.**
FIGURE **2.3**

▶ **A VORTEX FORMS IN A TORNADO.**
FIGURE **2.4**

Eye of hurricane

Land

Ocean

▶ **THE EYE OF A HURRICANE IS THE CENTER OF A LARGE VORTEX.**
FIGURE **2.5**

REFLECTING
ON WHAT
YOU'VE DONE

1 Think about what happened during your inquiry. Relate it to what happens during storms on the earth. Record your answers to these questions in your science notebook:

A. What happened when you first held the model so that the water was in the top bottle? Why do you think this happened? Draw a picture of your results.

B. How did you get the water to flow quickly into the bottom bottle?

C. How did the motion of the glitter and beads change as they moved closer to the center of the spiral?

D. Think about your model as a tornado. What might the glitter and beads represent? What does their movement tell you about the movement of air within a rotating storm?

E. Think about your model as a hurricane. What do you think causes the clouds of a hurricane to spiral?

2 Look carefully at Figures 2.3, 2.4, and 2.5. Using your experience with the vortex formed in your bottle, develop a working definition for the word "vortex." Write your definition in your science notebook. Then discuss it with the class.

3 Read "Tornado Watch or Warning?" on pages 24–25. Discuss the following questions with your classmates:

A. Why are weather alert warnings helpful?

B. Where might a weather alert and emergency alert warning system be most helpful?

4 Look again at your group's concept map and world map from Lesson 1. Focus on what you already know about storms. What do you want to add? What do you want to change? Use markers and dots, respectively, to revise or update your concept map and world map.

What Is a Vortex?

What happened when you used the vortex model in this lesson? You swirled water inside a bottle, letting gravity pull the water through the bottleneck. At the same time, air in the bottom bottle moved up through the center of the swirling water. This kind of circulation is called a vortex.

Vortices (the plural of "vortex") often form in nature. Swirling leaves, waterspouts, dust devils such as the one shown at right, and tornadoes are all examples of vortices. A vortex can also form in a hurricane that covers a very large

▶ IN THE DRY LANDSCAPE OF ARIZONA, A DUST DEVIL SWIRLS BROWN DUST UPWARD. A DUST DEVIL IS A KIND OF VORTEX.
PHOTO: NASA

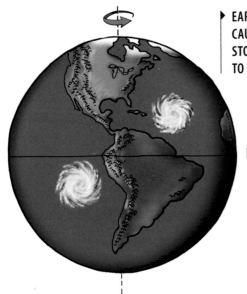

EARTH'S ROTATION CAUSES SOME STORMS AND WINDS TO ROTATE.

Equator

geographic area. The eye, or calm region at the center of a hurricane, is an example of the center of a large vortex.

What causes vortices to form in the atmosphere? The uneven heating of Earth's surfaces, the force of gravity, and Earth's rotation all influence air's motion and can force masses of air to rotate.

The rotation of Earth is one of the factors responsible for the formation of hurricanes. You can see how by trying a demonstration with a partner. As the illustrations to the left show, first lay a piece of paper on a table. Hold a pen in the center of the paper. Slowly move the pen's tip toward the edge of the paper and draw a straight line. Now have your partner rotate the paper counterclockwise, mimicking Earth's rotation, as you try to draw a line from the center of the page to its edge. You'll find that the line curves to the right.

Likewise, as air travels across Earth, Earth's rotation may curve its path. In later lessons, you will learn more about how air moves and what causes the vortex of a storm to form. ■

A STRAIGHT LINE WILL CURVE WHEN THE PAPER ROTATES.

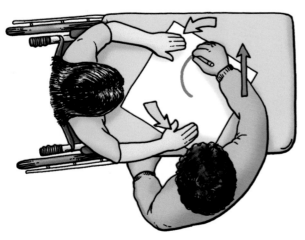

 DISCUSSION QUESTIONS

1. What are some examples of vortices that you have seen in real life or in photos?

2. How do you think a vortex is produced by the uneven heating of the earth's surfaces, the force of gravity, and Earth's rotation?

READING SELECTION

EXTENDING YOUR KNOWLEDGE

THAT'S A FACT: An Introduction to

THUNDERSTORMS, TORNADOES, and Hurricanes

What do these three pictures show? They show different kinds of storms, but can you tell which is a thunderstorm, which is a tornado, and which is a hurricane? Are there any clues that help you decide? Read on to learn about some of the characteristics of each kind of storm.

All three types of storms have a few things in common. They all involve rain and high winds, and each usually occurs at certain times of the year in certain locations. They can be forecast, detected, and tracked by the National Weather Service.

▶ PHOTOS: Top Left: NOAA Department of Commerce/NOAA Photo Library, NOAA Central Library; OAR/ERL/National Severe Storms Laboratory (NSSL), Bottom Left: NOAA Department of Commerce/OAR/ERL/National Severe Storms Laboratory (NSSL), Right: NOAA Department of Commerce

THUNDERSTORMS

A thunderstorm is a disturbance in the atmosphere that is characterized by lightning and thunder. Lightning is an electrical discharge in the air generated by charged particles in moving air masses. Because lightning is a phenomenon of moving, charged particles, not rain, we see lightning in violent forest fires and volcanoes as well as thunderstorms. Thunder is the sound produced by the shock wave lightning generates. The air immediately around lightning is suddenly heated to high temperatures—up to 30,000°C (54,000°F)—and subjected to high pressure; it expands rapidly.

Thunderstorms may be accompanied by gusty winds, heavy rain, sleet, snow, or hail, or by no precipitation at all. A severe thunderstorm can also produce flash floods and tornadoes. Thunderstorms generally move in the direction of overhead winds or in the direction of humid, unstable conditions. Here are some other key traits of thunderstorms:

- Thunderstorms can occur in any location, but are generally in the mid-latitudes. In the southeastern United States, thunderstorms occur most often along the Gulf Coast, especially in Florida, which experiences afternoon thunderstorms regularly in summer.

- Most thunderstorms occur in the spring and summer months during the warmest part of the day when warm air is most likely to be in motion. Other thunderstorms, for example, in the Central Plains, may occur at night.

- A well-developed thunderstorm can cover an area as large as 8 to 16 square kilometers (5 to 10 square miles). You probably recognize these clouds as thunderheads, or cumulonimbus clouds, which are often seen during a thunderstorm.

- If the temperature in part of a thundercloud falls below freezing and winds are strong, the raindrops in the storm can develop into hail. Although it is rare, a thunderstorm can occur during winter and may have snow as precipitation. This storm is called a "thundersnow."

▶ **YOU MAY RECOGNIZE THESE CLOUDS AS CUMULONIMBUS CLOUDS, WHICH ARE OFTEN SEEN DURING A THUNDERSTORM.**

PHOTO: Ralph F. Kresge/NOAA/Department of Commerce

- In a fraction of a second, a typical lightning bolt can discharge as much energy as a medium-sized nuclear reactor can in the same amount of time, with currents of up to 160,000 amperes. (Electrical circuits in most buildings carry about 20 amperes.)

- At any given moment, an estimated 1500 to 2000 thunderstorms are occurring on Earth. These storms can trigger 6000 or more lightning flashes per minute.

- The sound from a thunderstorm travels much more slowly than the lightning flash it produces: 340 m/s (1115 ft/s) for sound at sea level compared with about 3.0×10^8 m/s (about 186,000 mi/s) for light. Consequently, an observer will see the flash of lightning long before hearing the thunder. The time difference between the lightning flash and the sound of thunder can be used, along with the speed of sound, to calculate one's distance from the storm.

▶ A TORNADO THAT HIT PRATT CITY, ALABAMA, CAUSED WIDESPREAD DAMAGE. TARPS WERE USED AFTER THE STORM TO PROTECT THE CONTENTS OF HOMES FROM FURTHER DAMAGE.

PHOTO: FEMA/Tim Burkitt

TORNADOES

How do tornadoes form? A tornado is a rotating column of air that forms from thunderstorms over dry land under special conditions: when moist, warm air meets cool, dry air head on. Hurricanes can also induce tornadoes. Some facts about tornadoes:

• The United States has more tornadoes than any other country in the world. Most tornadoes in North America happen in "Tornado Alley" in the western plains or in "Dixie Alley" in the lower Mississippi Valley between Texas and the Gulf states and Tennessee.

• About three-fourths of all tornadoes in the United States develop from March to July, during late afternoon. The month of May normally has the greatest number of tornadoes in the United States (averaging about five per day), while the most violent tornadoes seem to occur in April.

• The diameter of most tornadoes is between 100 and 600 meters (328 and 1969 feet), although some are just a few meters wide, and others are wider than 1600 meters (1 mile).

• Some tornadoes stand nearly still, while others move at speeds faster than 100 kilometers (62 miles) per hour. There is no way to predict the exact path of a tornado, though meteorologists can estimate it within a band several miles wide.

• A tornado's vortex contains swirling winds that can move up to 350 kilometers (217 miles) per hour.

- Waterspouts are tornadoes that form over water.
- Much of a tornado's destructive power comes from its strong winds, which can lift huge objects and turn them into dangerous, high-speed missiles.

Japanese-born meteorologist Tetsuya "Ted" Fujita, who was known as "Mr. Tornado," developed the Fujita scale for measuring tornadoes on the basis of the damage they cause. The original Fujita Scale was linear; the winds in an F1 tornado were about twice as fast as those in an F0 tornado. Those winds in an F2 tornado were about three times as fast as those in an F0 tornado.

In 2007, the Fujita scale was revised to what is called the Enhanced Fujita Scale (see the table titled "Enhanced Fujita Scale of Tornadoes"). Data

ENHANCED FUJITA SCALE OF TORNADOES

F-SCALE NUMBER	INTENSITY PHRASE	WIND SPEED	TYPE OF DAMAGE
F0	GALE TORNADO	105–137 kph (65–85 mph)	Tree branches broken and shallow-rooted trees pushed over. Covering on school roofs damaged. Shingles peeled off mobile homes. Light material blown, such as from service station canopies.
F1	WEAK TORNADO	138–178 kph (86–110 mph)	Trees uprooted. Flagpoles and light poles bent. School windows broken. Mobile homes rolled on their sides or upside down. Roof panels stripped from service station canopies. Covered walkways collapsed at strip malls. Moving automobiles pushed off roads.
F2	STRONG TORNADO	179–218 kph (111–135 mph)	Tree trunks snapped, trees uprooted. Flagpoles and light poles collapsed. Masonry walls (e.g. in gym or cafeteria) of schools collapsed, roof structures damaged. Mobile homes separated from undercarriages and blown away. Service station canopies collapsed as columns buckle. Roof decking and covering on townhouses damaged.
F3	SEVERE TORNADO	219–266 kph (136–165 mph)	Trees debarked and uprooted with only stubs of largest branches remaining. Exterior and interior walls of schools collapsed. Mobile homes completely destroyed. Most walls of top stories of townhouses collapsed and roof structures uplifted. Large sections of big box stores destroyed. Cars overturned.
F4	DEVASTATING TORNADO	267–322 kph (166–200 mph)	Large sections of school buildings destroyed. Top stories of townhouses totally destroyed. Inside and outside walls collapsed on single-family homes. High-rise buildings deformed structurally. Buildings with weak foundations blown some distance. Large objects including cars thrown.
F5	INCREDIBLE TORNADO	>322 kph (>200 mph)	Engineered and well-constructed buildings completely destroyed. Automobile-sized structures carried long distances. All trees downed with bark stripped off.

Source: Storm Prediction Center, NWS, NOAA

from tornado damage surveys was used to better understand what is likely to happen to structures during a tornado. While the Enhanced Fujita Scale uses the same design as the previous version with six categories for degree of damage, the new scale better relates wind speeds to the amount and type of damage likely to occur. Because it is based on actual data, there is little information on impacts of the most extreme category of Incredible Tornadoes because, thankfully, they are rare.

Contrast this Enhanced Fujita tornado speed scale to the Richter scale for earthquake intensity. The Richter scale is measured on a logarithmic scale. What this means is that for each whole number you go up on the Richter scale, the size of the ground motion recorded by a seismograph goes up by a power of 10. So, for example, an earthquake that measures 6 on the Richter scale is 10 times stronger than one that measures 5 (and is 100 times stronger than one that measures 4, and so on.)

▶ **DAMAGE IN GUATEMALA FROM HURRICANE STAN**

PHOTO: U.S. Agency for International Development (USAID)

HURRICANES

What distinguishes hurricanes from thunderstorms and tornadoes? One big difference is that hurricanes are massive rotating storms, formed when warm, moist air rises over tropical waters. Some hurricane facts:

• People have different names for these rotating storms in different parts of the world. For example, a rotating storm is called a hurricane when it forms north of the equator in the Atlantic and eastern Pacific Oceans. It is called a typhoon when it forms north of the equator in the western Pacific Ocean. It is called a cyclone when it forms in the Indian Ocean and off the coast of Australia.

SAFFIR/SIMPSON HURRICANE SCALE

CATEGORY	WIND SPEED	DAMAGE	TYPE OF DAMAGE	STORM SURGE*
1	119–153 kph (74–95 mph)	MINIMAL	No significant damage to buildings. Damage primarily to unanchored mobile homes, shrubbery, and trees. Some coastal flooding and minor pier damage.	1–2 meters (m) (4–5 feet [ft])
2	154–177 kph (96–110 mph)	MODERATE	Some damage to roofing, doors, and windows. Considerable damage to vegetation and mobile homes. Flooding damage to piers and small craft.	2–2.5 m (6–8 ft)
3	178–208 kph (111–129 mph)	EXTENSIVE	Some structural damage to small residences and utility buildings. Mobile homes destroyed. Flooding near coast destroys small structures; floating debris damages larger structures. Inland flooding possible.	2.7–3.7 m (9–12 ft)
4	209–251 kph (130–156 mph)	EXTREME	Complete roof structure failure on small homes. Major erosion of beaches. Inland flooding possible.	4–5.5 m (13–18 ft)
5	252+ kph (157+ mph)	CATASTROPHIC	Complete roof failure on many homes and industrial buildings. Some buildings completely destroyed. Small utility buildings blown over or away. Flooding causes major damage to lower floors of all structures near shoreline. Massive evacuation of people possible.	5.5+ m (18+ ft)

*An unusually high water level, primarily due to winds during a storm, especially a hurricane.
Source: National Hurricane Center, NWS, NOAA

• Hurricanes can vary in size. For example, Typhoon Tip in the northwest Pacific had a diameter of 1100 kilometers (684 miles), while Tropical Cyclone Tracy had a diameter of just 100 kilometers (62 miles) when it struck Australia.

• Hurricanes can move at speeds of 8 to 24 kilometers (5 to 15 miles) per hour. Sometimes they can become nearly stationary. In that case, they inflict even more damage in one location.

• Hurricanes begin as tropical storms, which have wind speeds of 64 to 118 kilometers (40 to 73 miles) per hour. Hurricanes have wind speeds of 119 to 250 kilometers (74 to 155 miles) per hour or more. U.S. engineer Herbert Saffir and meteorologist Bob Simpson developed a scale that uses wind speed and storm surge to categorize the damage likely from hurricanes. You can see the kind of damage associated with various wind speeds in the table titled "Saffir/Simpson Hurricane Scale" on page 35.

• A hurricane rarely becomes more powerful when it hits land because it gets most of its energy from the warm ocean water beneath it.

• Because hurricanes can last a week or longer and several can occur at the same time, naming them reduces confusion. Names are selected in advance by governments of each region. Compare the names selected by the U.S. with those selected by the Philippines (see the table titled "Selected Hurricane Names for Atlantic Ocean and Philippine Region, 2013-2017). ■

DISCUSSION QUESTIONS

1. If you had to suffer a thunderstorm, a hurricane, or a tornado, which would you choose? Explain why.

2. If you were to make a Fujita-type scale for thunderstorms, what would it look like? Design one.

SELECTED HURRICANE NAMES FOR ATLANTIC OCEAN AND PHILIPPINE REGION, 2013–2017

	2013	2014	2015	2016	2017
Atlantic Ocean	Andrea	Arthur	Ana	Alex	Arlene
	Barry	Bertha	Bill	Bonnie	Bret
	Chantal	Cristobal	Claudette	Colin	Cindy
	Dorian	Dolly	Danny	Danielle	Don
	Erin	Edouard	Erika	Earl	Emily
	Fernand	Fay	Fred	Fiona	Franklin
	Gabrielle	Gonzalo	Grace	Gaston	Gert
	Humberto	Hanna	Henri	Hermine	Harvey
	Ingrid	Isaias	Ida	Ian	Irene
	Jerry	Josephine	Joaquin	Julia	Jose
	Karen	Kyle	Kate	Karl	Katia
	Lorenzo	Laura	Larry	Lisa	Lee
	Melissa	Marco	Mindy	Matthew	Maria
	Nestor	Nana	Nicholas	Nicole	Nate
	Olga	Omar	Odette	Otto	Ophelia
	Pablo	Paulette	Peter	Paula	Philippe
	Rebekah	Rene	Rose	Richard	Rina
	Sebastien	Sally	Sam	Shary	Sean
	Tanya	Teddy	Teresa	Tobias	Tammy
	Van	Vicky	Victor	Virginie	Vince
	Wendy	Wilfred	Wanda	Walter	Whitney
Philippine Region	Auring	Agaton	Amang	Ambo	Auring
	Bising	Basyang	Bebeng	Butchoy	Bising
	Crising	Caloy	Chedeng	Cosme	Crising
	Dante	Domeng	Dodong	Dindo	Dante
	Emong	Ester	Egay	Enteng	Emong
	Feria	Florita	Falcon	Frank	Feria
	Gorio	Gloria	Goring	Gener	Gorio
	Huaning	Henry	Hanna	Helen	Huaning
	Isang	Inday	Ineng	Igme	Isang
	Jolina	Juan	Juaning	Julian	Jolina
	Kiko	Katring	Kabayan	Karen	Kiko
	Labuyo	Luis	Lando	Lawin	Labuyo
	Maring	Milenyo	Mina	Marce	Maring
	Nando	Neneng	Nonoy	Nina	Nando
	Ondoy	Ompong	Onyok	Ofel	Ondoy
	Pepeng	Paeng	Pedring	Pablo	Pepeng
	Quedan	Queenie	Quiel	Quinta	Quedan
	Ramil	Reming	Ramon	Rolly	Ramil
	Santi	Seniang	Sendong	Siony	Santi
	Tino	Tomas	Tisoy	Tonyo	Tino
	Undang	Usman	Ursula	Unding	Undang
	Vinta	Venus	Viring	Violeta	Vinta
	Wilma	Waldo	Weng	Winnie	Wilma
	Yolanda	Yayang	Yoyoy	Yoyong	Yolanda
	Zoraida	Zeny	Zigzag	Zosimo	Zoraida

Note: Names are maintained by the World Meteorological Organization for the Atlantic Ocean and the Philippine Atmospheric Geophysical and Astronomical Services Administration for the Philippine Region.

HEATING EARTH'S SURFACES

▶ HOT CONCRETE AND COOL WATER ARE SIGNS THAT THE EARTH'S SURFACES HEAT AND COOL AT DIFFERENT RATES.

PHOTO: daveynin/
creativecommons.org

INTRODUCTION

Have you ever walked barefoot on a sidewalk in the early summer? The concrete probably felt hot against your feet. But if you jumped into a pool on the same day, you might have felt cold. How could this be?

Part of the explanation has to do with the way the earth's surfaces receive and give off heat. All of the surfaces on the earth absorb some of the sun's energy and give off heat to the air as they cool—but they do it at different rates. Did you know that the earth's surfaces heat and cool differently? In this lesson, you will investigate rates at which soil and water heat and cool. In later lessons, you will see that this uneven heating affects the circulation of air on the earth and helps create storms.

OBJECTIVES FOR THIS LESSON

Brainstorm ways to investigate how soil and water heat and cool.

Observe and record the rates at which equal volumes of soil and water heat and cool.

Graph and analyze the heating and cooling rates of soil and water.

Interpret data to compare ocean and land temperatures.

Explain what happens to energy from the sun when it reaches the earth.

Describe the atmosphere and its layers.

▶ MATERIALS FOR LESSON 3

For you

1 completed copy of Student Sheet 2.1: Thunderstorms, Tornadoes, and Hurricanes

1 copy of Inquiry Master 3: Graph Paper

1 copy of Student Sheet 3.1a: Testing the Heating and Cooling Rates of Soil and Water

1 copy of Student Sheet 3.1b: Interpreting a Data Table

1 metric ruler

1 blue pencil or pen

1 red pencil or pen

For your group

1 plastic box with lid

2 beakers

2 digital thermometers

2 cardboard strips

1 clamp lamp with bulb

2 bookends

1 stopwatch

1 transparency of Inquiry Master 3: Graph Paper

2 transparency markers (red and blue)

Soil

Water

Access to electricity

GETTING STARTED

1 Your teacher will review your homework with the class. Following the questions on Student Sheet 2.1: Thunderstorms, Tornadoes, and Hurricanes, discuss what you know about thunderstorms; where typhoons, hurricanes, and cyclones are likely to form; and the similarities and differences between tornadoes and hurricanes.

2 Brainstorm with your class ways you might investigate how soil and water heat and cool.

▶ **GATHERING STORM CLOUDS AND HIGH WINDS PROMPT A TORNADO WARNING IN NORTH DAKOTA.**
FIGURE **3.1**
PHOTO: FEMA/Patsy Lynch

INVESTIGATING RATES OF HEATING AND COOLING

PROCEDURE

1 Review with your teacher how to use a stopwatch and digital thermometer.

2 As a class, go over Procedure Steps 5 through 13. Observe as your teacher demonstrates the steps needed to complete the investigation.

3 With your class, review the Safety Tips for this inquiry.

SAFETY TIPS

Keep water away from all electrical outlets.

Avoid touching the hot lamp during the investigation and while the lamp is cooling.

Tuck electrical cords beneath work areas.

Do not drape cords across traffic areas.

Be careful with the sharp end of the thermometers.

Inquiry 3.1 continued

4 Review Student Sheet 3.1a: Testing the Heating and Cooling Rates of Soil and Water as your teacher discusses it.

5 Think about this investigation as if it were a test between equal volumes of soil and water. Which factors or variables are you changing? Which ones are you keeping the same? Under Step 1 on the student sheet, record all the things you think you will need to keep the same for both setups.

6 Pick up your group's materials. To set up a fair test, get equal volumes of soil and water from the distribution center.

7 Set up the materials as shown in Figure 3.1. Make sure that each beaker is the same distance from the heat source. (Why is it important to do this?) Insert each thermometer approximately 2.5 centimeters (cm) into the soil or water in each beaker. Do not allow the tip of the thermometer to touch the bottom of the beaker. Use the small hole in the cardboard to hold each thermometer upright. Turn on the thermometers.

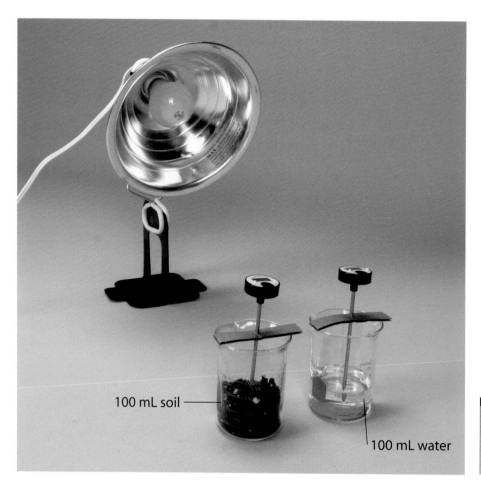

100 mL soil

100 mL water

▶ **SETTING UP THE INVESTIGATION. HOW WILL YOU MAKE CERTAIN IT IS A FAIR TEST?**
FIGURE **3.2**
PHOTO: ©2011 Carolina Biological Supply Company

8 Allow the thermometers to sit in each beaker until the temperature readings no longer show any sign of changing.

9 While you are waiting for the temperature readings to stop changing, make some predictions. What do you think will happen to the temperature of the soil and the water when you turn on the lamp? What will happen when you turn off the lamp? Why do you think this? Record your predictions under Step 2 on Student Sheet 3.1a.

10 Do not turn on the lamp yet. After the readings on the thermometers have stabilized, record the temperatures for both the soil and the water in Table 1 on Student Sheet 3.1a, across from 0:00 minutes under the columns labeled "Heating."

11 Turn on the lamp.

12 Start your stopwatch. Read the temperature of both materials to the nearest 0.1°C every minute for 10 minutes. Record your data in the table.

13 At the end of 10 minutes, turn off your lamp but let the watch keep running. Quickly record the 10-minute temperature for soil and water in the Heating columns. Record the same number across from 10:00 minutes at the top of the Cooling columns. Continue reading and recording the cooling temperature for soil and water every minute for 10 minutes.

Inquiry 3.1 continued

14 When you finish, clean up by doing the following:

A. Turn off the digital thermometers.

B. Dispose of the water from your beaker in a sink or bucket.

C. Do not throw away the soil. Pour it into the empty container set out by your teacher, where it can cool completely.

D. Return your materials to your plastic box for the next class.

E. Place your lamp in the spot selected by your teacher.

15 Complete Student Sheet 3.1a. Calculate the overall change in temperature of each beaker during heating and cooling. For the Heating columns, subtract the first temperature (at 0:00 minutes) from the last temperature (at 10:00 minutes). For the Cooling columns, subtract the last temperature (at 20:00 minutes) from the first temperature (at 10:00 minutes). Give your answers to the nearest 0.1 of a degree.

16 How might you plot your data on a graph so it is easy to read? Before you begin, discuss your ideas with your class.

READING SELECTION

BUILDING YOUR UNDERSTANDING

WEATHER VERSUS CLIMATE

In Part 1 of *Understanding Weather and Climate*, the words "weather" and "climate" appear quite often. Although both terms have to do with the state of the atmosphere, they do not mean exactly the same thing. How are they different?

Weather is the state of the atmosphere at a particular time and place. For example, a friend may ask, "What was the weather like when you were at the beach yesterday?" Your friend wants to know about the weather on a certain day at a certain place. Your answer might be, "In the morning it was rainy, but then it got really hot and windy."

Your answer refers to three important elements of weather: moisture, temperature, and wind speed. (Another element is atmospheric pressure.) Remember that weather is the state of the atmosphere. So, where there is no atmosphere, there is no weather like that on the earth. This means that beyond the earth's atmosphere in space, there is no weather as we know it.

Climate refers to weather patterns that are characteristic of a region or place for many years. For example, you might describe the southwestern region of the United States as having a hot, dry climate. Or you might say that the climate of a town near the coast is more moderate than the climate of a town 100 kilometers inland. Over many thousands of years, the climate can change, and this has happened in most parts of the world during long geologic spans of time. Later in the unit you will investigate the methods used by scientists to collect data that is associated with climate change. ■

17 Work with your group to create one graph to share with the class. While you do, consider these questions:

A. What title will you give your graph?

B. How will you label each axis to show the temperature and time changes?

C. What will be the first number on each axis? How will you space the numbers on each axis? How many units (minutes or degrees) will each interval between the numbers represent?

D. What techniques will you use to make the graph more readable?

18 Compare your group's completed graph with another group's graph. Analyze them using the questions in Procedure Step 17. Do they look the same? Is one easier to interpret than the other? What accounts for the similarities and differences between the graphs?

19 On your own, graph your group's data using Inquiry Master 3: Graph Paper and the tips you discussed with your teacher and the class. Title your graph and follow the suggestions discussed in Steps 16 and 17.

REFLECTING
ON WHAT
YOU'VE DONE

1 Answer and then discuss these questions with the class:

A. How would you describe the heating and cooling rates of soil and water in this investigation?

B. Which material held its heat longer?

C. What factors may have influenced your results?

D. Reread the Introduction to this lesson. Can you explain now why concrete feels hot under your feet in early summer, while water in a pool feels cold?

E. Based on your investigation, how do you think oceans absorb and hold heat? How do you think the temperature of the ocean compares with the temperature of the land nearby?

2 Discuss "The Source of Earth's Heat," on pages 46–49, which you read for homework, with your class. Review the questions at the end of the reading selection.

3 Complete Student Sheet 3.1b: Interpreting a Data Table. You should also read "The Atmosphere: A Multilayered Blanket of Air," on pages 50–52.

4 Read "Weather Versus Climate." Look ahead to Lessons 4 and 5. In these lessons, you will investigate how the uneven heating of the earth's surfaces affects weather in the earth's atmosphere.

The SOURCE of EARTH'S HEAT

After school, the air is warm, moist, and breezy; fat raindrops hit the pavement. Lightning flashes, and thunder rumbles in the distance. In the west, the sky is dark. A thunderstorm is headed your way. What causes the weather and its storms? They are caused in part by the effects of heat from the Sun.

SOLAR ENERGY

Solar energy is the source of most of Earth's heat on land, in the oceans, and in the atmosphere. (The rest comes from the decay of radioactive elements in Earth's core.) Solar energy radiates through the vacuum of space to Earth in the form of electromagnetic waves. Some of this solar radiation is visible as light, and some (heat, for example) is invisible. The interactions between solar energy and air, soil, and water on Earth creates wind, rain, and other elements of weather.

That sounds simple enough, right? Wrong! Only a tiny fraction of the Sun's energy strikes Earth. Of the small amount that does reach Earth, about half is absorbed by the planet's surface. The rest is reflected back into space or absorbed by the thin blanket of air—the atmosphere—that surrounds Earth.

Earth has various kinds of surfaces that respond differently to the Sun's energy. Materials such as soil, rock, and water absorb and give off energy at different rates. Look at the illustration at right. Snow's surface absorbs only 5 percent of the solar energy it receives, while a dark forest absorbs 95 percent of the solar energy it receives. The temperature of a surface, such as snow or a forest, is an indication of the amount of the Sun's heat energy that has been absorbed. Differences in how Earth's surfaces absorb and give off energy relate to wind and weather patterns. Because

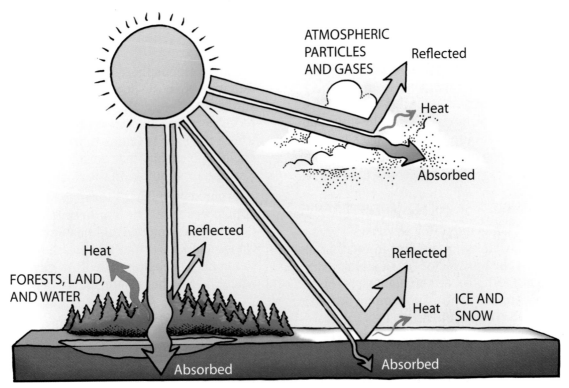

ATMOSPHERIC PARTICLES AND GASES

Reflected

Heat

Absorbed

Reflected

Heat

FORESTS, LAND, AND WATER

Reflected

Heat

ICE AND SNOW

Absorbed

Absorbed

the planet is unevenly heated, air rises, expands, cools, and sinks at different rates in different places. These differences in air temperature and pressure cause movements of the air—from areas of high pressure to low, from warm areas to cool—and this is what we know as wind.

All the heat reflected from Earth's surfaces would travel back into space unimpeded and leave the planet cold if it were not for the atmosphere. Certain gases in Earth's atmosphere absorb some of the heat energy, trapping it in the atmosphere and then sending it back toward the surface again. These gases—primarily water vapor, carbon dioxide, methane, ozone, and nitrous oxide—are called greenhouse gases, and their return of heat energy to Earth's surface is called the greenhouse effect. It is reminiscent of the way glass traps the Sun's heat in a greenhouse and warms the air inside, sometimes to the point of

▶ EARTH'S ATMOSPHERE AND SURFACES ABSORB AND REFLECT THE SUN'S ENERGY. SOME OF THE ABSORBED ENERGY IS GIVEN OFF AS HEAT.

READING SELECTION
EXTENDING YOUR KNOWLEDGE

being stifling hot. In the last few centuries, the concentration of greenhouse gases in Earth's atmosphere has increased, and scientists have observed Earth warming. Later in the unit, you will study more about climate changes in the past and the present, and about those predicted for the future.

SEASONS MAKE A SIGNIFICANT DIFFERENCE

The ways air, land, and water absorb and give off energy play an important role in weather. But these are not the only factors in manufacturing the weather. Earth, like all the planets, revolves around the Sun. Earth also spins on its own axis, which is tilted. The tilt of Earth's axis barely changes, but the part of the planet that gets the most direct and strongest solar energy changes with the seasons as Earth orbits the Sun.

Many people believe that seasons change as Earth moves toward and away from the Sun in its elliptical orbit. That might seem logical, but consider this: at certain times, Earth is slightly closer to the Sun in December than in June. If proximity to the Sun were responsible for Earth's temperature, this would mean that everyone on Earth would have summer in December. You know this is not true if you live in the Northern Hemisphere.

It is the tilt of Earth's axis that is responsible for changes in seasons. From about March 21 through September 21, the Northern Hemisphere is tilted toward the Sun. During this time, the Northern Hemisphere receives more direct solar radiation than the Southern Hemisphere. When sunlight falls directly on a surface, that surface receives more solar energy than does a surface where the light is falling more indirectly, at an angle. This means that during this time it is warmer, or summer, in the North.

From September 23 through March 19, the opposite is true. The Southern Hemisphere is tilted toward the Sun and has warmer weather. On December 21, for example, the Southern Hemisphere celebrates the first day of summer, while the Northern Hemisphere begins winter.

On two days of the year (around March 21 and September 21), neither hemisphere is tilted toward the Sun. Therefore, both hemispheres receive the same amount of the Sun's energy. These two days are called the equinox.

A DELICATE BALANCE

Without energy from the Sun, all the things that we take for granted on Earth, including the weather, would not exist. Weather distributes heat and precipitation (such as rain) around the globe. Gases and clouds in the atmosphere hold in the amount of heat needed to keep Earth livable. The atmosphere affects the amount of solar energy that reaches Earth and protects it from the Sun's more harmful radiation. The atmosphere and its weather keep most of Earth's surfaces warm enough for life to exist as we know it.

Life survives within a narrow range of temperatures; the survival range for individual species is narrower still. Currently, Earth is neither too warm (like Venus or Mercury) nor too cold (like Neptune). However, Earth's temperature has not always been and may not always be hospitable to life. This is why climate scientists have a genuine sense of urgency about monitoring and caring for the atmosphere, to keep the planet livable for humans and other species. ■

DISCUSSION QUESTIONS

1. Why is it hot in summer and cold in winter?

2. A reader has proposed a solution to global warming: the "earth parasol," a sunshade that would be suspended between Earth and the Sun. Assuming it were possible to make and deploy such a sunshade, do you think it would work? Why or why not? What unintended effects might it have?

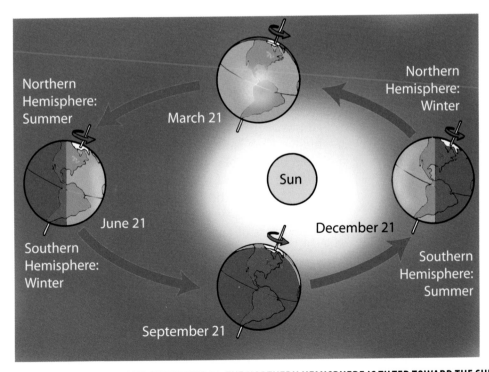

▶ BETWEEN MARCH 21 AND SEPTEMBER 21, THE NORTHERN HEMISPHERE IS TILTED TOWARD THE SUN AND HAS SPRING AND SUMMER. DURING THAT SAME TIME, THE SOUTHERN HEMISPHERE IS TILTED AWAY FROM THE SUN AND HAS FALL AND WINTER. THE EQUATOR IS WARM ALL YEAR ROUND. (DIAGRAM NOT DRAWN TO SCALE.)

THE ATMOSPHERE: A MULTILAYERED BLANKET OF AIR

What makes Earth distinct from all the other planets in our solar system? Its atmosphere! The atmosphere acts like a blanket of air around Earth, holding in heat. It also reflects some of the solar energy that reaches Earth, protecting us from overheating, and shields us from the Sun's more harmful radiation. The force of gravity holds Earth's atmosphere in place. The Moon has no atmosphere because its gravitational force is much smaller than that of Earth.

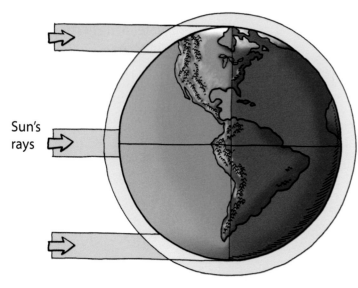

Sun's rays

▶ EARTH'S ATMOSPHERE ABSORBS AND REFLECTS SOLAR ENERGY. AT THE POLES, BECAUSE OF THE CURVE OF THE EARTH, SUNLIGHT HAS TO PASS THROUGH MORE ATMOSPHERE TO REACH THE EARTH'S SURFACE THAN THE SUNLIGHT THAT REACHES THE EQUATOR. THIS IS ONE REASON WHY THE POLES ARE COLDER THAN OTHER PARTS OF THE EARTH.

SLICES OF SKY

Think of Earth's atmosphere as multiple layers above its surface. (See the illustration on the next page.) The bottom layer, the troposphere, is where most of Earth's weather takes place. The surface of Earth influences the temperature, moisture, and wind velocity (speed and direction) in the troposphere. This layer contains most of the water vapor in the atmosphere.

Air moves in all directions in the troposphere—up, down, and sideways. This is due to the uneven heating created by radiation from the Sun above and from the land and oceans below. (In Lessons 5 and 6, you will investigate what happens to air when it is heated by the surface below it.) The temperature in the troposphere usually gets colder as one goes higher. This is because the troposphere is heated from below. Solar radiation strikes the land and the seas and warms them, and they release heat back into the atmosphere. The part of the atmosphere closest to the heat-releasing Earth is the bottom of the troposphere.

Above the troposphere is the stratosphere. The stratosphere protects Earth from the Sun's harmful radiation. Here the temperature rises with altitude. Few gases are present here, but one of them is quite important to life on Earth: ozone. Ozone (O_3) is a form of oxygen. It is an almost colorless gas with an odor similar to weak chlorine and can be detected in violent thunderstorms in the troposphere. Ozone

	Exosphere	Exosphere extends out to about 1000 km; most satellites orbit here.
500 km		
	Thermosphere	Northern lights occur here.

Space stations such as MIR and the International Space Station have stable orbits at 320–380-km altitude.

Temperatures here may reach as high as 1000°C. |
280 km		
		The ionosphere exists in the lower thermosphere between 80 and 280 km.
80 km	Mesosphere	Temperatures decrease here to −90°C; meteoroids burn up here.
50 km	Stratosphere	Very little turbulence; ozone layers absorb radiation here.
15 km	Troposphere	Contains 78% nitrogen, 21% oxygen

▶ SCIENTISTS DIVIDE THE EARTH'S ATMOSPHERE INTO FIVE LAYERS.

collects in the middle of the stratosphere and forms a protective layer, trapping the Sun's harmful ultraviolet radiation and keeping it from reaching the troposphere. Its ability to trap solar radiation is responsible for the warmer temperatures at the top of the stratosphere.

Over the poles of Earth are holes in this protective layer of ozone. We managed to create these holes in the ozone layer ourselves by using refrigerant and spray chemicals called chlorofluorocarbons, or CFCs. CFCs are quite safe for humans and other creatures; not so safe, unfortunately, for the ozone layer. CFCs were produced for decades in the 20th century, and when they escaped into the atmosphere, they reacted with ozone. CFC production has been stopped, and the chemicals have been replaced with other fluorocarbons that do not deplete ozone nearly as much.

The third layer of the atmosphere is the mesosphere. At the top of this layer, the temperature decreases to about minus 90°C (minus 130°F) before reaching the final layers: the thermosphere and exosphere.

The thermosphere is where solar radiation first meets the atmosphere, and where extreme ultraviolet rays are absorbed. Temperatures here can go very high: over 1700°C (3100°F). The thermosphere actually expands and shrinks depending on the amount of solar radiation it's absorbing, but in 2010 a puzzling and alarming thing happened: the thermosphere appeared to collapse. It has since expanded again, but atmospheric scientists have no explanation for the dramatic contraction, and are continuing to monitor the thermosphere.

There is very little air in both the thermosphere and the exosphere. Satellites, other spacecraft, and meteors can travel through the thermosphere and exosphere with very little resistance. In the exosphere, Earth's atmosphere fades into the vacuum of outer space. ■

DISCUSSION QUESTIONS

1. We live in the troposphere. Could we live in any of the other layers of the atmosphere? Why or why not?

2. What problem do chlorofluorocarbons cause for life on Earth?

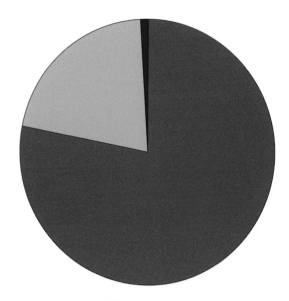

■ Nitrogen: 78%
■ Oxygen: 21%
■ Trace gases, including water vapor: about 1%

▶ **NITROGEN AND OXYGEN MAKE UP THE MAJORITY OF ATMOSPHERIC GASES.**

JOSEPH HENRY:
The Father of
Weather
Forecasting

▶ **JOSEPH HENRY**

PHOTO: Library of Congress, Prints & Photographs Division, LC-BH824-4499

Your team has a soccer game Saturday, so you check a local news station's website to see the weekend forecast. Radar images on multicolored maps show rain moving east, away from your town, and bands of clouds a few hundred miles west. It could mean rain, but the forecast for Saturday is partly cloudy with a high of 75 degrees. You check a national weather site instead, hoping for something more precise, and check the hour-by-hour forecast. Saturday, 10 AM: partly sunny and 68 degrees. Perfect.

In this age of 10-day weather forecasts and colorful digital displays of the entire country's weather, it's hard to imagine not being able to find out tomorrow's forecast. But before the mid-1800s, farmers and ship captains, whose lives and jobs depended on the weather, had little information to go on. They relied on clouds, winds, the Farmer's Almanac, past experience in how the seasons flow, animal behavior signs, and their own arthritic bones to make predictions about the weather. But a scientist named Joseph Henry changed all of that.

If you have ever heard of Joseph Henry, it was probably in connection with the Smithsonian Institution, the world's largest museum and research complex, located in Washington, D.C. Joseph Henry was the Smithsonian's first director. He and his family lived in the first building of the Smithsonian—the Castle—when that building was in a rural setting, before a bustling capital city grew up around it.

READING SELECTION
EXTENDING YOUR KNOWLEDGE

▶ THE TELEGRAPH WAS THE NEW TECHNOLOGY OF THE DAY. WITH A TAP OF A TELEGRAPH KEY, OPERATORS COULD SEND WEATHER INFORMATION TO THE SMITHSONIAN INSTITUTION.

PHOTO: NSRC/Anne Williams

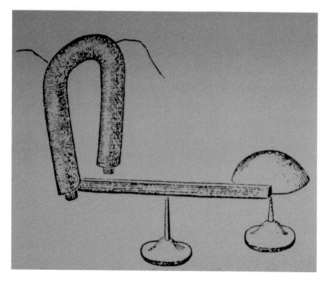

▶ A SKETCH OF THE "TELEGRAPH" HENRY SHOWED HIS CLASSES AT THE ALBANY ACADEMY. ALTHOUGH PEOPLE CREDIT SAMUEL B. MORSE WITH INVENTING THE TELEGRAPH, HENRY WAS ACTUALLY TINKERING WITH IT YEARS BEFORE.

PHOTO: Smithsonian Institution Archives. Image SIA2012-7651.

SPREADING THE METEOROLOGICAL WORD

From the start, Henry was determined to keep this national museum's focus on science, and a key science in his mind was meteorology. "Of late years, in our country," he wrote, "more additions have been made to meteorology than to any other branch of physical science." For this reason, Henry decided to develop a "system of extended meteorological observations." Observing the weather nationally, he thought, could solve "the problem of American storms."

Like any good scientist, Henry developed a plan of action. First, he needed weather information from around the country. He established a network of about 150 volunteer observers, a number that eventually grew to about 600. The Smithsonian supplied the volunteers with instructions, standardized forms, and, in some cases, instruments. The volunteers submitted monthly weather reports from their area. This included temperature, humidity, wind and cloud conditions, and rain and snow measurements. Analyzing this information required years of study, but eventually it helped scientists to better understand storms, weather patterns, and climate differences across the country.

Weather forecasting was another part of Henry's plan. He arranged for nearly 20 telegraph stations around the country to report weather information to the Smithsonian once a day. Compared with the monthly reports from volunteer observers, the information from telegraph stations was not detailed. Telegraph operators communicated only whether the sky was clear or cloudy, whether it was raining or snowing, and the direction of the wind.

Henry posted this information on a large map in a public area of the Smithsonian Institution (see the picture on page 55). He put white discs on cities with clear skies, blue on those with snow, black on those with rain, and brown on those with cloudy skies. Under Henry's direction, the Smithsonian Institution had

A SCIENTIST WITH AN UNCONVENTIONAL EDUCATION

Although Joseph Henry was a famous and honored American scientist in his day, his origins were humble. Born into a poor family in New York, Henry had to leave school for financial reasons after he finished his elementary years. He went to work as an apprentice to a silversmith and watchmaker. Henry's life changed, however, when he read George Gregory's *Lectures on Experimental Philosophy, Astronomy, and Chemistry*. At the age of 22 he resumed his formal education at Albany Academy and made such progress that he frequently taught the other students and his teachers, as well.

Henry later served as a professor at the New Jersey college that became Princeton University. He was a lifelong learner with interest in the natural and physical sciences who continued to read and teach himself science and mathematics. Weather was just one of the subjects that fascinated Joseph Henry. In fact, his studies of electromagnetism laid the foundation for the development of the telegraph and telephone, and a unit of measurement used in electromagnetism is named the "henry" in his honor.

He encouraged other scientists, such as Alexander Graham Bell, who were interested in sound transmission and room acoustics. He inspired Thaddeus Lowe's work in aeronautics for researching weather conditions in the atmosphere and the jet stream. After his work at the Smithsonian, Joseph Henry became the second president of the National Academy of Sciences, the most prestigious group of scientists in America, and used his position and influence to promote the study of science throughout the country.

▶ JOSEPH HENRY'S WEATHER MAP WAS PROBABLY THE FIRST ONE IN THE COUNTRY.

PHOTO: Smithsonian Institution Archives. Image 84-2074.

READING SELECTION

assembled, for the first time, "one view of the meteorological condition of the atmosphere over the whole country." Although crude by today's standards, the map attracted public attention and prompted discussions about the need for a national weather service.

Henry had solved the problem of getting weather information to the Smithsonian Institution, but data by itself is useless. It must be organized and then analyzed if anything is to be learned from it. So Henry found a colleague, James H. Coffin, who organized the weather information into reports. In 1861, Coffin published two volumes of weather information collected from 1854 to 1859.

METEOROLOGICAL DATA LEADS TO NEW DISCOVERIES

With so much weather data now available in useful form, scientists made new discoveries and developed new theories about weather. Henry theorized that local storms are part of larger weather systems. Another scientist, Increase A. Lapham, used the data to show that a storm moved across the country from west to east and that the path could be plotted on a map. This finding meant that communities could be warned about storms moving their way. Henry quickly realized that the telegraph could be a part of such an early-warning system. In this way, people in the eastern parts of the country could be warned of storms well ahead of time.

WHAT HAPPENED TO THE SMITHSONIAN'S WEATHER NETWORK?

In 1857, Henry planned to expand the weather network he had set up into a storm warning system for the East Coast. But the Civil War soon overwhelmed the country and its telegraph lines. When the war was finally over, Henry suggested that the federal government establish a national weather service.

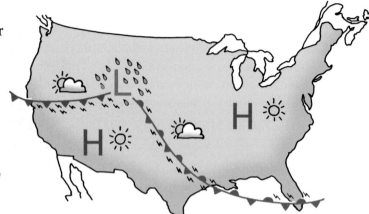

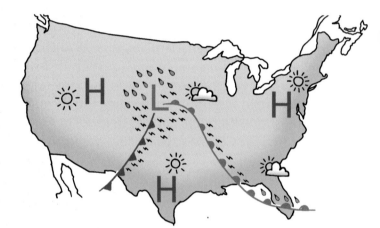

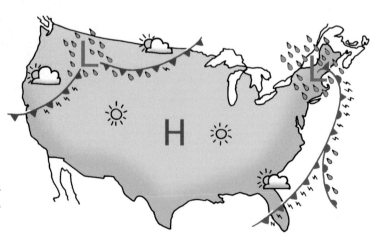

▶ **WEATHER SYSTEMS FOLLOW WIND PATTERNS AND MOVE ACROSS THE UNITED STATES FROM WEST TO EAST.**

Today's weather service is highly advanced compared with Henry's system of telegraphed observations and a simple wall map, and yet in some ways not much has changed. Henry's visions of monitoring weather on a large scale and developing a science of weather have been realized in the sophisticated forecasting system we have today. He is considered by many to be the "father of modern weather forecasting."

Although Henry recognized that it would probably be impossible to achieve perfection in weather forecasting, accuracy in weather forecasting has improved dramatically since his time. The use of satellites and other sophisticated tools and instruments developed over the last 50 years has enabled forecasters to deliver more precise predictions with greater confidence than Henry could have imagined. ■

DISCUSSION QUESTIONS

1. Henry provided instruments to some of his weather-watchers. If you were running his data-collection mission, would you allow people to use their own instruments? Why or why not?

2. If Joseph Henry were alive today, what would he think of our weather monitoring systems? What would surprise him and what would not?

HEAT TRANSFER AND THE MOVEMENT OF AIR

> **HANG GLIDERS STAY ALOFT BY TAKING ADVANTAGE OF RISING CURRENTS OF AIR.**
>
> PHOTO: Nick Hewson/ creativecommons.org

INTRODUCTION

How can a hang glider stay up in the air for hours without a motor? How can a bird soar over an open field without flapping its wings? They can do these things largely because of heat transfer and the motion of air. As the wings of a glider or a bird lift it up in the air, gravity pulls it down. This means that glider pilots and soaring birds need to find constant upward forces to stay in the air. Do you know where these forces come from? They come from huge masses of warm, rising air. Flat fields, dark pavement, and low-lying towns absorb a great deal of heat early in the day. These materials also give off large amounts of heat. Where there is a warm surface, you are sure to find warm, rising air.

In this lesson, you will investigate what happens to air when it is heated or cooled by the surface beneath it. How does heat move between the earth's surface and the air above it? How do surface temperatures on the earth affect the temperature of air above it and the way air moves? You will investigate these ideas in Lesson 4. Then, in Lesson 5, you will look at what happens when air masses of different temperatures meet.

OBJECTIVES FOR THIS LESSON

Compare land and water temperature at different times during a 24-hour period.

Investigate the effect of surface temperature on the temperature of air above the surface and its resulting movement.

Hypothesize about the transfer of heat from the earth's surface to the air above it.

Determine the basic conditions under which water moves through the air.

Develop working definitions for the words "stable air mass" and "unstable air mass."

▶ **MATERIALS FOR LESSON 4**

For you

1	completed copy of Student Sheet 3.1b: Interpreting a Data Table
1	copy of Student Sheet 4.1: Investigating the Temperature of Air

For your group

1	plastic box with lid
2	Convection Tubes™
1	120-mL plastic container of hot water (with screwtop lid)
1	120-mL plastic container of crushed ice (with screw-top lid)
1	stopwatch
1	digital thermometer
1	metric ruler
2	rubber bands
2	pieces of plastic wrap
1	piece of plastic tubing
1	small funnel
1	punk stick
1	flashlight
1	small aluminum pan
1	pair of scissors
	Paper towels

GETTING STARTED

1 Go over your homework from Lesson 3, Student Sheet 3.1b: Interpreting a Data Table, with your teacher. As you do, think about the following:

A. What was the temperature of Portland Parklands at 2:00 P.M.?

B. How does this temperature compare with the temperature of the Atlantic Ocean near Portland, Maine, at 2:00 P.M.?

C. How do you think the temperature of the earth's surface affects the temperature of air above it?

D. How do you think the temperature of air affects how air moves?

Discuss your ideas with the class.

2 In this lesson, you will investigate how the temperature of a surface affects air temperature, air movement, and the formation of clouds. Your teacher will show you a Convection Tube™. Brainstorm with your class ways in which you might use the tube to explore this interaction.

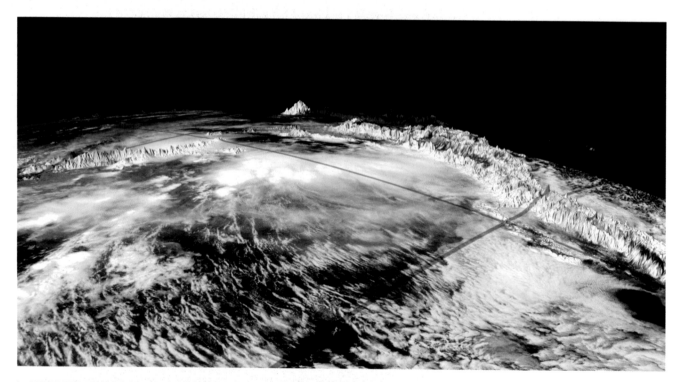

▶ **SATELLITE IMAGE OF FIELDS OF CIRRUS CLOUDS ON TOP OF A TROPICAL STORM IN CENTRAL AMERICA**

PHOTO: NASA/Goddard Space Flight Center Scientific Visualization Studio. The Next Generation Blue Marble data is courtesy of Reto Stockli (NASA/GSFC). MODIS data courtesy of Jeff Schmaltz, MODIS Rapid Response Project (NASA/GSFC).

INVESTIGATING THE TEMPERATURE OF AIR

PROCEDURE

1 Look over Student Sheet 4.1: Investigating the Temperature of Air as your teacher discusses it. Read the question at the top of the student sheet. You will complete the student sheet as you conduct Inquiry 4.1.

2 Observe as your teacher demonstrates the setup and Steps 5 through 12 of the Procedure. Review Figures 4.1 and 4.2 with your teacher at this time.

3 How would you make this investigation a fair test? List your ideas under Step 1 on Student Sheet 4.1.

4 Make a prediction of the temperature of the air in the tube above different surfaces, then record it under Step 2 on the student sheet. Discuss your prediction with your class.

5 Collect your materials. With your group, practice reading the thermometers inside the cylinders. The number on the thermometer highlighted with green is the correct temperature. If two numbers that are not green are highlighted, you can average them. Also practice reading the time on the stopwatch and starting and stopping it until you feel comfortable using it.

6 If it has not been done for you, fill one plastic container with hot water and one with crushed ice.

7 Use the digital metal thermometer to measure the temperature of the hot water. Also measure the temperature of the crushed ice. Write the temperatures for ice and hot water in the second row of Table 1 on the student sheet.

Inquiry 4.1 continued

8 Before you place each container of water (without its lid) under a Convection Tube™, record the starting temperatures of both thermometers in both cylinders on Table 1 of the student sheet. Write them across from Time 0:00. (Thermometer A is the top thermometer.)

9 Set your stopwatch at zero. Place the container of hot water under one Convection Tube. Place the container of crushed ice under the other Convection Tube, as shown in Figure 4.1. Then start the stopwatch.

10 Record the changes in temperature in each Convection Tube every minute for three minutes. If the temperature goes higher than the thermometer's highest temperature, you can record 30+°C on your data table. (Do not touch the outside of the cylinder. Your hand may affect the temperature readings.)

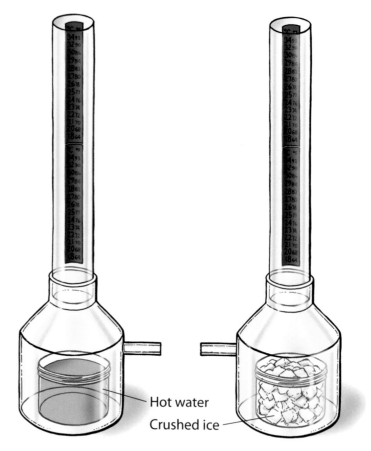

Hot water

Crushed ice

▸ PLACE THE HOT WATER AND CRUSHED ICE UNDERNEATH THE CONVECTION TUBES.
FIGURE **4.1**

11 If it gets difficult to see inside the Convection Tube, use a paper towel to remove moisture from the base. Attach a paper towel to a ruler with a rubber band and use this device to clear the cylinder and base, as shown in Figure 4.2. After clearing the Convection Tube, you can cover your containers of water or ice with plastic wrap and secure the wrap with a rubber band.

12 Clean up your work area.

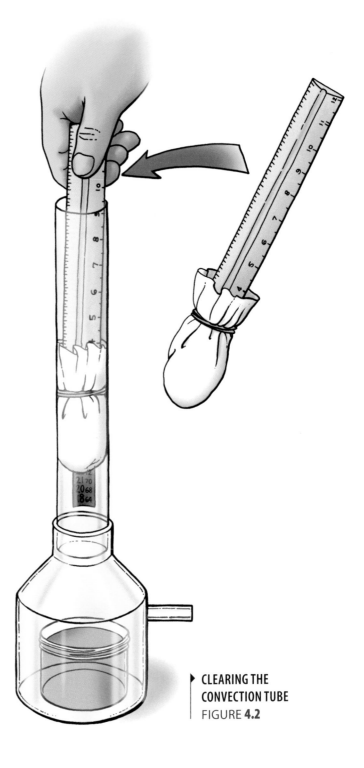

▶ **CLEARING THE CONVECTION TUBE**
FIGURE **4.2**

INQUIRY 4.2

INVESTIGATING HOW WARM AIR AND COOL AIR MOVE

PROCEDURE

1 Record the purpose of this investigation in your science notebook. Write it in the form of a question. Then share it with your group or class. Also share with the class your ideas on how to test this question. 🖉

2 Your teacher will demonstrate how to do this investigation. Follow along using Procedure Steps 4 through 9.

3 Make a prediction about the temperature of the air in the cylinder. How do you think air will move above a hot surface? How do you think air will move above a cold surface? Record your predictions. Discuss your ideas with your group or class.

4 Pick up your materials. Set up the Convection Tubes™ with hot water and crushed ice, as you did in Inquiry 4.1. You will not be recording temperature changes in this investigation. Use the ruler and paper towel to clear the cylinder or temporarily cover the containers with plastic wrap.

5 Attach the funnel and tubing to the Convection Tube with ice. (It is important that you begin with the ice.)

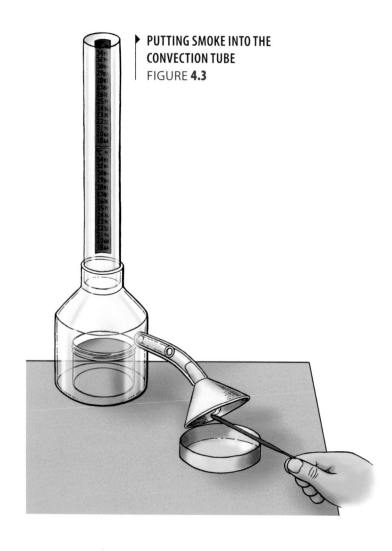

▶ PUTTING SMOKE INTO THE CONVECTION TUBE
FIGURE **4.3**

6 When you are ready, ask your teacher to light your group's punk stick. Immediately blow it out and hold the smoking punk over the aluminum pan, as shown in Figure 4.3. Position the funnel at an angle over the punk so the smoke goes inside. Do not touch the funnel with the burning punk.

SAFETY TIP

Follow safety precautions when working with a burning punk. Do not walk around the room with the punk while it is burning.

7 Your teacher will turn off the classroom lights. Use your flashlight to see the smoke particles moving. Hold the flashlight behind the Convection Tube and then at the top of it. Do not cover the opening of the Convection Tube. (See Figure 4.4.) Kneel down so you can see the smoke at eye level as it enters the Convection Tube. Talk to your partners about how the smoke moves.

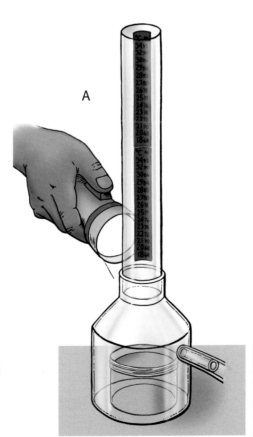
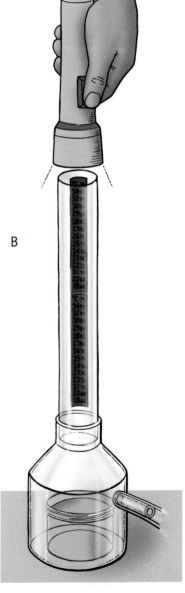

A B

▶ (A) SHINE THE FLASHLIGHT FROM BEHIND THE CONVECTION TUBE TO VIEW THE AIR AS IT ENTERS THE CYLINDER. (B) HOLD THE FLASHLIGHT AT THE TOP OF THE CONVECTION TUBE TO SEE THE SMOKE THROUGHOUT THE CYLINDER. DO NOT BLOCK THE OPENING OF THE CYLINDER.
FIGURE **4.4**

Inquiry 4.2 continued

8 Move the tubing and funnel to the Convection Tube™ with hot water. (One member of your group should carefully continue to hold the burning punk.) Clear the cylinder with the ruler and paper towel if needed. Place the punk under the funnel to add smoke to the Convection Tube with hot water. Observe. Use the flashlight to view the smoke.

READING SELECTION

BUILDING YOUR UNDERSTANDING

AIR MASSES

When air moves over different surfaces—for example, cold mountains or a warm ocean—it takes on the temperature and humidity (moisture) conditions of that area. Because of this, air separates into massive pockets, or air masses. An air mass has the same temperature and moisture content throughout. It can extend for hundreds or thousands of kilometers.

Once formed, air masses can move and carry their weather conditions to another area. For example, air masses that come out of northern Canada are cold and dry. Air masses that form over cold oceans bring cold temperatures and moisture in the form of ice or snow. Air masses from the Gulf of Mexico that are warm and moist bring clouds and rain showers. Air masses from Texas, New Mexico, and Arizona that are warm and dry bring hot temperatures in the summer.

Where cold and warm air masses meet, a distinct boundary forms between them. The denser, cold air mass may slide under the warm one and lift it up. The weather at the boundary becomes unstable. When this happens, stormy weather may be ahead. ■

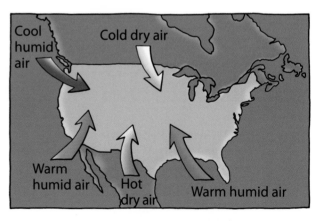

▶ **WHERE DO YOU THINK DIFFERENT AIR MASSES OFTEN MEET IN THE UNITED STATES? WHY?**

9 When you have finished observing the Convection Tube with hot water, clean up your work area. Carefully dip just the tip of the punk into a container of water. This will stop the tip from smoldering. Then cut off the wet tip.

REFLECTING
ON WHAT YOU'VE DONE

1 Answer these questions in your science notebook; then discuss them with the class:

A. Based on your temperature readings in Inquiry 4.1, how does the temperature of a surface affect the temperature of the air above it?

B. In Inquiry 4.2, how did the air, visible because of the smoke, move in the Convection Tube with crushed ice? Explain why you think this happened.

C. What happened to the air inside the Convection Tube with hot water? Explain why you think this happened.

D. Why do you think moisture formed on the inside of the Convection Tube with hot water? How do you think this relates to the process of cloud formation on earth?

E. Apply what you observed in this lesson to the earth. If the earth's surface is colder than the air above it, what will happen to the air? If the surface is hot and damp, what will happen to the air above it?

2 Read "Air Masses."

3 With your teacher's help, develop working definitions for "stable air mass" and "unstable air mass." Record your definitions in your science notebook. Apply what you observed in this lesson to cloud formation. When do you think clouds are more likely to form: when air remains close to the earth's surface or when air rises and moves quickly to high altitudes?

4 Look ahead to Lesson 5, in which you will connect two Convection Tubes to investigate what happens when air masses meet.

READING SELECTION
EXTENDING YOUR KNOWLEDGE

What's the Weather Forecast?

▸ **BOB RYAN, METEOROLOGIST AND WEATHER FORECASTER**

PHOTO: Bob Ryan, Chief Meteorologist NBC 4 (WRC), Washington, D.C.

What will the weather be like today? To find out, we may turn on the television to see what weather forecasters such as Bob Ryan have to say.

"An important part of my job," says Ryan, "is to make the forecast as clear and understandable as possible." Whether he is forecasting a blizzard, severe thunderstorms, or a sunny spring day, he says, "It comes down to hand-holding with people because they're concerned. They ask: 'Can I go out? What should I wear?'" His responsibility is to give people as much useful information as possible.

RESEARCHING THE WEATHER

Ryan is more than the reporter in front of the camera. As a meteorologist, Ryan is a scientist with access to more than 3500 weather stations, and with their data, he makes his own predictions. Every hour, these stations provide information on air temperature, air pressure, wind direction and speed, relative humidity, and precipitation.

Ryan also uses Doppler radar to detect how air is moving. How does Doppler radar work? Think of the sound an ambulance makes as it approaches. The pitch of the siren gets higher as it approaches you; that's because the sound waves bunch up, and have a higher frequency, just ahead of the ambulance. As it passes you and moves away, the pitch gets lower: the sound waves stretch out, and the frequency drops.

Doppler radar works something like that. A device sends out microwaves at a particular frequency to bounce off a target, such as an air mass or storm cloud. If the reflected frequency is higher than the original, a target is moving toward the radar. If the frequency is lower, the cloud or air mass is moving away from the radar. The magnitude of the frequency shift also reveals how fast the masses are moving. Doppler radar has enabled forecasters like Ryan to provide people with early warnings of potential danger from storms.

Why is radar so important to the study of thunderstorms, high winds, and tornadoes? Radar is important primarily because these storms cannot be seen from satellites, and they can develop very quickly. Seeing them form and approach gives meteorologists a chance to warn people in their path.

▶ DOPPLER RADAR TOWER

PHOTO: NOAA/ Department of Commerce/NOAA Photo Library, NOAA Central Library; OAR/ERL/ National Severe Storms Laboratory (NSSL)

READING SELECTION

Satellites are also important tools for Ryan. Satellites orbiting in space take pictures of clouds covering the earth. A series of pictures over time can show meteorologists how large storms like hurricanes move, and can help predict the paths they might take.

In addition to radar and satellite information, Ryan also uses data from ground weather instruments (for example, thermometers, anemometers, and barometers). Even special airplanes collect data that are mapped and fed into computer models. The computer models calculate wind, precipitation, temperature, and weather movement at locations around the globe. Ryan compares predictions calculated by several different models to see how well they agree.

Ryan says that the science of weather forecasting is a lot like the science of medicine. Medical doctors use all the available tests, X-rays, scans, and other methods to diagnose a patient's condition. Meteorologists do the same thing to arrive at a forecast, he says. They ask: "What are the weather data and the calculations showing? What does it mean to our local area?"

PRESENTING THE WEATHER

When Ryan presents the weather, he decides what part of the local weather is most important to viewers. "Each situation is a little bit different," he says. If thunderstorms are forming in the Blue Ridge mountains west of Washington, he shows the radar pattern. "If I'm following a tropical system, like a hurricane, I might use almost hourly satellite images," he explains.

Ryan often links his weather report to other news. For instance, he once showed a tropical storm forming over the Pacific Ocean. He explained how the storm's rainfall could help put out fires that were raging in Mexico. He is always interested in following storm forecasts (snow, rain, ice, etc.) that could affect his viewers' lives and presents long-range as well as short-range forecasts.

▶ BOB RYAN GIVES A FORECAST IN FRONT OF A GREEN
BOARD CALLED A CHROMA-KEY. TELEVISION VIEWERS
SEE WHAT IS SHOWN ON THE TOP TV MONITOR.

PHOTO: Bob Ryan, Chief Meteorologist NBC 4 (WRC), Washington, D.C.

▶ BOB RYAN WORKING IN THE CHANNEL 4
STORM CENTER

PHOTO: Bob Ryan, Chief Meteorologist NBC 4 (WRC),
Washington, D.C.

READING SELECTION
EXTENDING YOUR KNOWLEDGE

INCREASING ACCURACY OF FORECASTS

With the advent of satellites and other technology, weather forecasts have become increasingly accurate. The accuracy of most weather predictions is directly related to the length of the forecast. A five-day forecast will be more accurate than a ten-day forecast; a short-term six-hour forecast, called nowcasting, can accurately predict the time, intensity, and length of individual rain showers or storms. Most forecasts are based on complex models that use the different variables that affect the weather. Scientists on the East Coast of the United States can use information from weather tracking stations across the country to make more precise predictions of weather moving eastward. Scientists on the West Coast, however, do not have the same advantage in tracking weather across the ocean before it reaches land.

Forecaster Bob Ryan and his peers, however, have reached a level of weather prediction accuracy and sophistication that would probably amaze and gratify their predecessor, Joseph Henry. How accurate are they? For the month of August 2010, the accuracy of 5-day weather forecasts for Washington, D.C., from five forecasters varied from 60% correct to 84% correct. Now, you may think that's not impressive; after all, at the low end, the forecasters got the weather forecast right little more than half the time. Consider, though, that a forecaster is predicting what will happen five days away, when the weather systems that will affect the area are still hundreds or thousands of miles away. That would have been inconceivable a hundred years ago. As the forecast date gets nearer, accuracy goes up substantially.

INSPIRING THE NEXT GENERATION

Some years ago, Ryan decided he wanted to get more young people excited about the weather. He created a program called 4-WINDS to teach students about weather. With the help of area businesses, NBC-4 donated weather stations to local schools. Each weather station contains instruments for measuring weather variables (including temperature, pressure, humidity, and wind) and a computer to store the data. He also visited classrooms to talk with students about weather and forecasting.

Ryan hopes that young people who are interested in the weather will have a better understanding of science and maybe even become meteorologists. They need to have a strong background in math and science and do well in those subjects, he says. Ryan learned about weather by studying science in college and earning a master's degree in physics and atmospheric science. "I think there's a lot of opportunity for people interested in this field—from improving short-term forecasts, to studying global climate change, to managing water resources. Most of all," he says, "it should be something you consider fun and look forward to doing." ■

DISCUSSION QUESTIONS

1. Ryan says that the science of weather forecasting is a lot like the science of medicine. How so?

2. How is finding patterns important in the study of weather?

WEATHER FORECASTING CAN BE COOL . . . OR HOT!

When Vicky participated in the 4-WINDS weather forecasting program with her classmates at Mountain View School in Haymarket, Virginia, she found that "it was like finding out how a magician does his tricks. They use computers to help them figure out the answers to the questions," she said. "I thought weather forecasting was a 'luck' thing. Maybe you got it right, or maybe you were way off." She found out that she could make predictions that were "pretty close" by looking at patterns of temperature, wind, and pressure.

"I used to think that computers were just for games and for typing," said Vicky's classmate, Shawn. "But there are other things you can do. I like using the 4-WINDS weather station because it lets me know what's going to happen. I'm getting better and better at figuring the weather out."

According to Amanda May, Vicky and Shawn's teacher, participating in the weather forecasting "gave the students a tremendous knowledge of how technology is used in everyday life. My students were getting very accurate weather forecasts for our area within 24 to 36 hours because they could track patterns. It really made an impression when they could see someone else calling our system over the Internet. And there we were on Channel 4! Or the National Weather Service would call [to get our data]. The students were doing adult work." Because of the weather program, May continued, several of her students developed a great interest in earth science and meteorology.

Carol is one such student. She wrote in her journal, "When the wind is from the southwest and the barometric pressure is falling, something is going to come out of the sky. Then we just use the thermometer to figure out what that something is, snow or rain. This weather stuff is a cinch!"

Shawn agreed. "I can predict what tomorrow's weather will be because I can find out what the weather is right now," she said. "I'm good at it, but I don't think I'm a threat to any forecaster...yet!"

CONVECTION CURRENTS IN THE AIR

> HOLD ON! A REALLY WINDY DAY IS GREAT FOR CERTAIN KINDS OF SPORTS.

PHOTO: Fiona Shields/creativecommons.org

INTRODUCTION

What causes the wind to blow? Mostly, it has to do with the uneven heating of the earth. Breezes occur because land and water heat and cool at different rates. When the sun's energy has heated the earth, the warm earth heats the air above it. Just above the ground, the air's temperature increases. The warming air begins to rise, and cool air moves in to take its place. In turn, the warm earth heats this air. Again, it rises, and cool air takes its place. These transfers of heat cause changes in the weather, including the development of winds.

In this lesson, you will connect two of the Convection Tubes™ you used in Lesson 4. What happens when two masses of air with the same temperature and humidity meet? What happens when air masses of different temperature and humidity conditions meet? After observing the movement of air in the convection model you set up, you will apply what you have learned to two real-world situations. You will analyze how land and sea breezes form, and how tornadoes develop.

OBJECTIVES FOR THIS LESSON

Set up an investigation that demonstrates what happens to two air masses when they meet.

Think about and analyze the origin, meeting, and movement of air masses with different temperature and humidity conditions.

Develop working definitions of the terms "convection current" and "weather front."

Relate the movement of air within the convection model to the formation of land and sea breezes and the development of tornadoes.

Explain how winds form.

▶ MATERIALS FOR LESSON 5

For you

1	copy of Student Sheet 5.1a: When Air Masses Meet
1	copy of Student Sheet 5.1b: Convection on the Earth

For your group

1	plastic box with lid
2	Convection Tubes™
1	piece of plastic tubing
1	flashlight
1	120-mL plastic container of hot water (with screwtop lid)
1	120-mL plastic container of crushed ice (with screw-top lid)
1	candle
1	punk stick
1	small aluminum pan
1	inflatable globe
1	pair of scissors
1	Weather and Climate World Map
1	set of multicolored dots

GETTING STARTED

1 Think back to Lesson 4. What did you discover about how the temperature of air affects the way air moves? Discuss this with your class.

2 Look back to the reading selection, "Air Masses" on page 66. Where do you think air masses with different temperature and humidity conditions are most likely to meet in the United States?

▶ THIS SATELLITE IMAGE OF NORTHERN AFRICA SHOWS DUST FROM A RECENT DUST STORM. THE MOVEMENT OF THE DUST REVEALS THE MOVEMENT OF THE AIR.

PHOTO: NASA image courtesy Jeff Schmaltz, MODIS Land Rapid Response Team at NASA GSFC

INVESTIGATING THE EFFECTS OF COLLIDING AIR MASSES

PROCEDURE

1 Obtain one copy of Student Sheet 5.1a: When Air Masses Meet. Read the question at the top of the student sheet. You will investigate this question during the inquiry.

2 Look at one set of connected Convection Tubes™ and the materials for each group. Then look at Table 1 on Student Sheet 5.1a. What are some ways you might set up this equipment to investigate the question in this inquiry? Discuss this with your class. One suggested setup is shown in Figure 5.1.

3 On your student sheet, make a list of the materials you will use and the procedures you will follow to test each setup. Be prepared to share your ideas with the class.

4 Which variables will you keep the same in each setup? Which variable(s) will you change during each test? Write down your ideas on Student Sheet 5.1a.

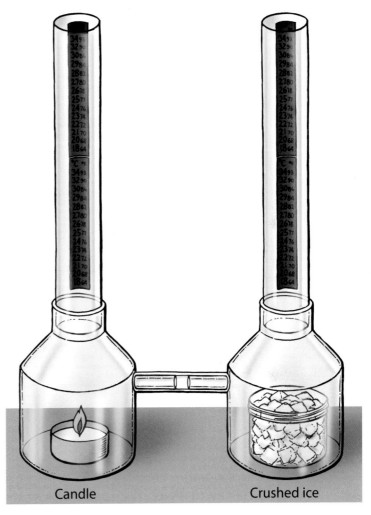

Candle   Crushed ice

▶ **CONNECTED CONVECTION TUBES**
FIGURE **5.1**

Inquiry 5.1 continued

5 What do you think will happen when cold, moist air meets cold, moist air? What do you think will happen when warm, moist air meets warm, moist air? What will happen when cold, moist air meets hot, dry air? Discuss your predictions with your group. Record what you think will happen in the Predictions column in Table 1 on your student sheet.

6 Review with your teacher the following points, which you should keep in mind while you work:

A. Keep the Convection Tubes™ connected at all times.

B. You will be making qualitative observations in this lesson, which means you will be watching and describing what happens, not taking measurements. You will not have to record any temperature changes.

C. Introduce smoke into the top of the tube, as shown in Figure 5.2.

7 Before you begin, review Safety Tips with your teacher.

8 Collect and set up your materials. Begin the investigation. Discuss your observations with your group as you work, and record them on your student sheet. For each setup, remember the procedures your group developed. Use your flashlight to see the smoke.

9 When you have finished testing all three conditions, clean up. Put out the burning punk by dipping just the tip of it in a cup of water. Cut off the wet tip with the scissors. Refill your container with crushed ice for the next class.

SAFETY TIPS

Roll up loose sleeves and tuck in loose clothing.

Tie back long hair.

Do not let the burning punk touch the tube. The plastic tube will melt if it does.

Do not ask your teacher to light your candle until you are ready.

Do not reach across an open flame.

Do not leave the candle under the plastic tube for longer than 1 minute. The plastic will get hot.

▶ USE THE PUNK STICK TO INTRODUCE SMOKE INTO THE TOP OF THE TUBE.
FIGURE **5.2**

REFLECTING
ON WHAT
YOU'VE DONE

1 Answer the following questions and discuss your observations with the class:

A. What did you observe when both tubes contained air with the same temperature and humidity conditions? Why do you think this happened?

B. What did you observe when the tubes contained air with different temperature and humidity conditions? Why do you think this happened?

C. Based on your results from Lessons 4 and 5, where and how do you think winds and rotating storms might form?

2 Look again at the illustration in "Air Masses" (page 66). Where in the United States do you think air masses with different temperature and humidity conditions might meet? The boundary that forms when this happens is called a weather front. What type of weather do you think might occur along a front?

3 A convection current formed when you set up the Convection Tubes so that a hot air mass collided with a cold one. Use your experiences to write your own definitions for the terms "convection current" and "weather front." Discuss your definitions with the class.

4 Read "Why Does the Wind Blow?" on pages 80–85 and "Weather Fronts" on pages 86–87. Revise your definitions if needed.

5 Your teacher will ask you to complete Student Sheet 5.1b: Convection on the Earth to find out what you know about how air moves when cooled and heated. On this sheet you will do these steps:

• Illustrate how air moved in your group's Convection Tube.

• Relate the movement of air within your convection model to the formation of land and sea breezes.

• Apply the movement of air within your convection model to the development of tornadoes.

6 Read and discuss "Danger in Dixie Alley" on pages 88–93.

7 Review the map and the globe used in Lesson 1 and your placement of the dots for tornadoes. Revise the placement of the orange dots based on your study of tornadoes.

Why Does the Wind Blow?

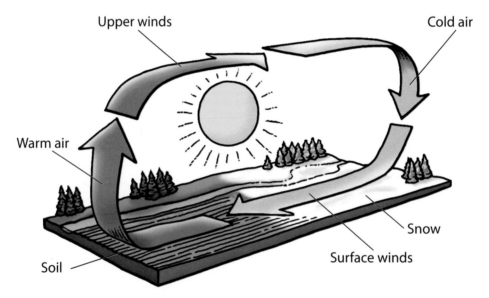

▶ UNEVEN HEATING CAUSES WINDS.

What makes the wind blow? From light breezes to strong gusts, winds are the result of uneven heating of the earth's surfaces.

The process begins as the sun warms the earth. As the layer of air above a warmed surface heats, it expands, becomes lighter, and rises. As the warm air rises, it expands and eventually cools. The dense, cold air moves in to replace the rising warm air. The earth warms this layer of incoming cool air and it too rises, and then is replaced by another layer of cooler air. This cycle goes on and on. The circulating flow of air resulting from temperature differences is called a convection current. Convection currents can form in liquids, too. Later in the unit, you will investigate convection currents in the ocean. Can you speculate about the causes of convection currents in the ocean?

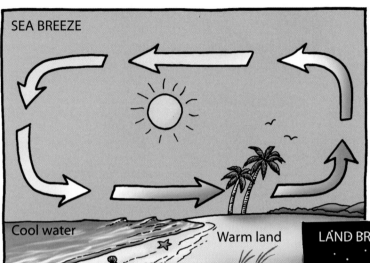

SEA BREEZE

Cool water

Warm land

► SEA BREEZES OCCUR DURING THE DAY.
LAND BREEZES OCCUR AT NIGHT.

LAND BREEZE

Warm water

Cool land

LAND BREEZES AND SEA BREEZES

Land breezes and sea breezes are both caused by convection, or the movement of heat through air or water. These breezes occur where large bodies of land and water meet. The different rates at which land and water heat and cool cause these winds.

During the day, land warms up faster than the water in lakes or oceans. Warm air rises above the land, forming an area of low pressure, and this allows the cool, denser air over the water to slide in and push aside the less dense warm air. This flow of air is called a sea breeze.

At night, the land cools faster than the water does, so the air over the land becomes cooler than the air over the water. As the warm air over the water rises, cool, dense air from the land moves toward the water and pushes warm, less dense air aside. The flow of air from land to water is called a land breeze.

Other factors can cause local winds. One factor is color. You may already know, for example, that dark-colored clothing absorbs heat faster than light-colored clothing. This same phenomenon affects surface heating on the earth, too. Dark soils absorb heat faster than light-green fields. The air above these surfaces is warmed or cooled accordingly, and local winds result.

READING SELECTION

EXTENDING YOUR KNOWLEDGE

GLOBAL WINDS

Other winds don't just blow locally. They are continually forming around the earth, moving in a particular direction—always to the east, for instance—and traveling over long distances. These winds, which form between the equator and the poles, are called global winds.

The equator and the poles are not heated evenly. Near the equator, the sun's rays are more direct and more intense. (See the illustration on page 50 in Lesson 3.) This is what makes the tropics so warm. Near the poles, sunlight falls on the surface at an angle, and passes through more atmosphere to reach the surface. Since the atmosphere both absorbs and reflects the sun's heat, less light hits each square meter of surface at the poles than at the equator. The result is that the poles are not as warm as the tropics. Recently, however, scientists discovered that the polar regions are absorbing heat at an increasing rate as the earth's climate changes, particularly as sea ice melts and sunlight strikes dark ocean

rather than white ice. This change is already having a significant effect on the ecosystems and living things in the polar regions, as you will find out in Lessons 10 and 12.

CIRCLING THE GLOBE

Global winds are the result of giant convection currents that circulate within the Northern and Southern Hemispheres of the earth. As warm air is heated at the equator, it rises and flows both north and south toward the poles. If the earth didn't rotate, the hot air at the equator would rise to the poles, cool, sink, and flow back to the equator again. (See the illustration below.) But the earth does rotate, which means that air and water currents on the earth are deflected.

Trade winds, westerlies, and easterlies are names of different kinds of global winds that form because of the earth's rotation and the sun's energy. Trade winds flow toward the equator, turning west as they go. Westerlies flow from west to east. These winds are called

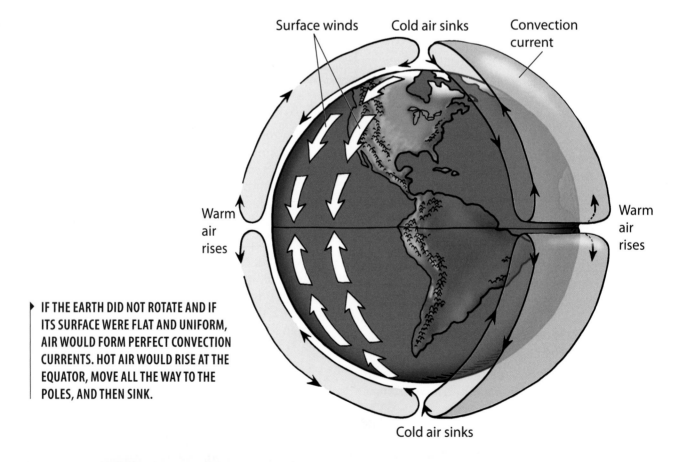

▸ IF THE EARTH DID NOT ROTATE AND IF ITS SURFACE WERE FLAT AND UNIFORM, AIR WOULD FORM PERFECT CONVECTION CURRENTS. HOT AIR WOULD RISE AT THE EQUATOR, MOVE ALL THE WAY TO THE POLES, AND THEN SINK.

WHAT ARE MONSOONS?

Monsoons are very powerful land and sea breezes that change direction with the seasons. They occur mostly in southern Asia and are an important part of life there.

During the summer, the air over the land heats up. As the hot air rises, it is replaced by warm, moist air from the Indian Ocean. In turn, this moist air is heated, and rises. The water vapor in the rising air condenses, forms clouds, and produces rain. The cycle repeats throughout the season, bringing long periods of rain to the region. During the summer, people grow rice and other crops that need a lot of moisture.

In the winter, the land cools faster than the water does. Cool air over the land sinks and moves out to sea. During this time of year, southern Asia receives little rain.

▶ **WOMEN MAKING THEIR WAY THROUGH THE FLOODING FROM A MONSOON IN SRI LANKA**

PHOTO: Hafiz Issadeen/creativecommons.org

westerlies because they flow *from* the west. For example, polar westerlies are winds that flow toward the poles, turning east as they go.

Easterlies flow from east to west. These winds are called easterlies because they flow *from* the east. For example, polar easterlies are winds that sink at the poles, spread outward, and turn west as they go. What happens when these global winds meet? Where the westerlies and easterlies meet, weather changes occur. The meeting of the westerlies and easterlies has a major effect on the weather that occurs in North America.

JET STREAM

Strong winds in what is called the jet stream are long, relatively narrow "tubes" of air in the boundary between the troposphere and stratosphere. They form at various latitudes from the equator to the poles, where the earth's curvature produces cooler zones next to relatively warm zones. Located about 10 km (6 mi) above the ground, and meandering in their paths, the jet streams are only a few hundred kilometers wide, but they sometimes stretch halfway around the earth.

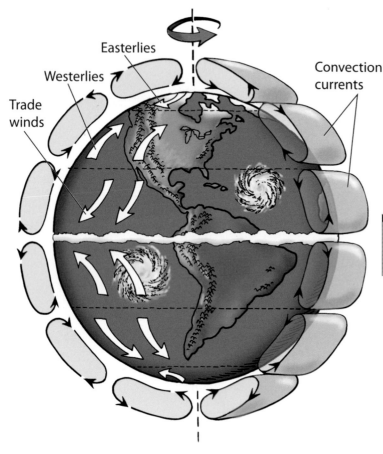

▶ THE EARTH'S ROTATION AND OTHER FACTORS CAUSE SMALLER CONVECTION CURRENTS TO FORM ON THE EARTH. THE RESULTING WINDS IN THE NORTHERN AND SOUTHERN HEMISPHERES ARE MIRROR IMAGES OF EACH OTHER.

Jet stream winds usually travel at about 200 kilometers (124 miles) per hour, but can move twice that fast. They were discovered by American pilots during World War II, although some 19th-century scientists speculated about their existence. Jet streams and their paths vary from day to day and season to season. These powerful winds play a large part in moving air masses around the earth, which means that they also play a big part in determining the earth's weather. Forecasters can use the path of a jet stream to predict how weather will move across the country. ■

DISCUSSION QUESTIONS

1. What causes all winds?

2. As warm, moist air forms over the ocean, prevailing winds will drive it toward land. When such air masses encounter a large mountain range, we observe a phenomenon called the rain shadow effect. As a result of the rain shadow effect, one side of the mountain will experience much more precipitation than the other. Can you predict which side of the mountain (that facing the ocean, known as the windward side, or that facing the land, known as the leeward side) gets more rain? Explain your prediction.

CLOUDS IN THE JET STREAM OVER THE MIDDLE EAST. THE JET STREAM HELPS HIGH-FLYING AIRPLANES TRAVEL EAST. PLANES GOING WEST TRY TO AVOID THE JET STREAM. CAN YOU FIGURE OUT WHY?

PHOTO: NASA

weather fronts

Have you ever heard your local weather forecaster talk about weather fronts? Fronts bring changes in the weather. They occur when air masses of different temperature, pressure, and humidity conditions collide. A weather front forms along this boundary between different air masses.

There are several types of fronts. A cold front is the leading edge of a moving mass of cold air. When a cold air mass pushes a warm air mass ahead of it, the dense, cold air slides under the light, warm air, just as cold water will sink to the bottom of a tub of hot water. The warm air gets pushed upward. If there is a lot of water vapor in the rising warm air, dense cumulonimbus clouds form, and rain or snow may fall. If the rushing upward of moist air masses is violent enough, there may be lightning (scientists are still not exactly sure why lightning occurs, though they do know that positive and negative charges separate high in the storm cloud). If there is little water vapor, clouds form, but don't result in storms. Cold fronts frequently move fast and cause abrupt changes in weather, including violent thunderstorms or tornadoes. After a cold front passes, cool, dry air moves in.

At a warm front, a moving, warm air mass overrides a cold air mass ahead of it. The warm air is less dense, so it slides over the cold air. If the warm air is dry, scattered clouds form. If the warm air is humid, rain (or light rain or snow in the winter) normally falls along the front. Warm fronts typically move slowly, so rainy weather usually stays around for days.

If two air masses move close to each other but neither has enough force to move the other, they both remain fixed in place. The boundary between them is called a stationary front. At the point where the warm air and cold air meet, water vapor in the warm air condenses into rain, snow, fog, or clouds. If the stationary front remains in place for a long time, it may bring days of clouds and precipitation.

In a more complex frontal system—an occluded front—a fast-moving cold air mass slides under a warm one, and then encounters another air mass. If this third air mass is warmer than the cold air, it's lifted up, trapped between the first warm air mass and the fast-moving cold air. If the third air mass is colder than the fast-moving cold front, the fast-moving air is the one to get sandwiched between two air masses. Either way, the sandwiched air mass is cut off, or occluded, from the ground. Unsettled, cloudy weather is associated with occluded fronts, including long periods of thunderstorms.

When you listen to weather forecasts from now on, pay close attention to what the forecaster says about fronts. Do you notice that one type of front tends to form in your area more than others? What type of weather and cloud cover does each front bring? ■

▶ WARM FRONT

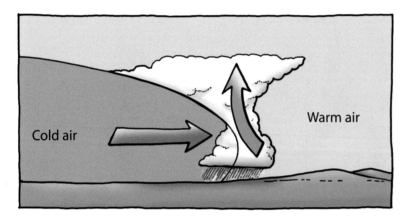

▶ COLD FRONT

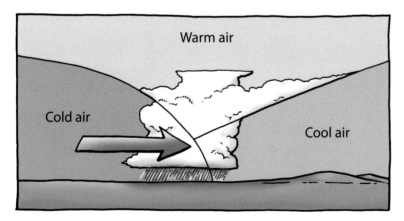

▶ OCCLUDED FRONT

DISCUSSION QUESTIONS

1. Why was the name "occluded front" given to a particular weather system? Consider the definitions of each word.

2. Write a fictitious weather forecast that includes at least one weather front.

DANGER in Dixie Alley

▶ THIS BUILDING WAS DESTROYED BY A DEADLY TORNADO THAT SWEPT THROUGH YAZOO CITY, MISSISSIPPI, IN APRIL 2010.

PHOTO: FEMA/George Armstrong

People living in the part of the United States called Tornado Alley are all too familiar with the springtime sound of a tornado siren. This area, which includes several states in the Great Plains region, is an ideal site for tornado formation. It's where dry polar air from Canada meets warm, moist tropical air from the Gulf of Mexico. Recently, however, scientists have focused their attention on another area of high tornado frequency and intensity: Dixie Alley, which includes several states near the Gulf Coast.

In 2010, Mike Frates, a 24-year old graduate student from The Ohio State University, mapped out all F3, F4, and F5 tornadoes (based on the Fujita scale, or F-scale) occurring in the United States over a 56-year period. He counted those that stayed on the ground for 20 minutes or more and grouped them into zones or "alleys." He found that the alley which he called Dixie Alley had more intense, long-travelling tornadoes than the famous Tornado Alley of the Plains states. One reason may be that tornado season in that area lasts all year long while it's a danger in Tornado Alley only about four months out of the year.

Frates reported that his research on Dixie Alley tornadoes took on new meaning in the spring of 2010 when a series of 88 tornadoes struck Louisiana, Arkansas, Mississippi, Alabama, and Georgia in a three-day period. Six of them were deadly F3 and F4 tornadoes. Typical of an F4 tornado was the one roaring

up from Louisiana that struck Yazoo City, Mississippi, around noon on April 24th. The storm's path, at some points, was 2.8 kilometers (1.7 miles) wide. As it traveled along its 240-kilometer (150-mile) track, the storm destroyed homes, stores, churches, businesses, and schools. With peak winds of 274 kilometers (170 miles) per hour, the tornado in Mississippi killed 10 and injured 146 people statewide.

What is it like to be in the path of one of these powerful storms? Yazoo City survivors told their stories to the press. As the tornado approached, a restaurant owner and his employees ran to the frozen foods locker. After ten minutes that "seemed like two hours," they

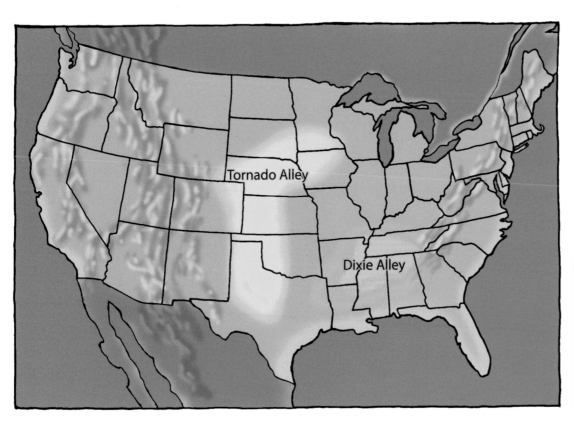

▶ WHILE TORNADO ALLEY IS MORE FAMOUS AS AN AREA OF HIGH TORNADO FREQUENCY, DIXIE ALLEY SUFFERS MORE SEVERE TORNADOES.

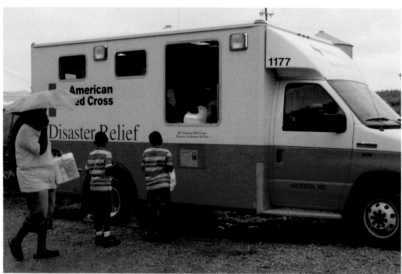

▶ THIS YAZOO CITY SURVIVOR IS STARTING THE DIFFICULT WORK OF REBUILDING AFTER THE DEVASTATING TORNADO.

PHOTO: FEMA/George Armstrong

▶ THE AMERICAN RED CROSS BROUGHT IN FOOD SUPPLIES TO DONATE TO FAMILIES WHO WERE IMPACTED BY THE TORNADOES.

PHOTO: FEMA/George Armstrong

STAYING SAFE

Where should a person go if a tornado strikes? A storm shelter or basement is the safest place to be, but even in Tornado Alley or Dixie Alley, many people don't have access to them. In that case, people try to find interior rooms with no windows, such as hallways and closets under stairwells. Some families climb into their bathtubs and cover themselves with blankets and cushions to protect against falling or flying debris.

To help people stay safe in weather emergencies such as tornadoes, relief workers from agencies such as the Federal Emergency Management Agency and the American Red Cross recommend that people stock their homes with the following supplies:

- FLASHLIGHTS AND EXTRA BATTERIES
- PORTABLE, BATTERY-OPERATED RADIO WITH EXTRA BATTERIES
- FIRST-AID KIT
- EMERGENCY FOOD AND WATER
- ESSENTIAL MEDICINE
- CHARGED CELL PHONE

▶ THIS STORM CELLAR SAVED THE LIVES OF A FAMILY WHO TOOK SHELTER IN IT WHILE THEIR HOUSE WAS DEVASTATED BY A TORNADO.

PHOTO: FEMA/Win Henderson

escaped unharmed to find the restaurant in shambles. The county coroner was driving in his pickup truck when the tornado crossed his path, flipping his truck and rolling it over several times. After the tornado moved on, he emerged from the truck and, despite his cuts and bruises, managed to help other injured survivors. A couple survived the demolition of their trailer home by hiding in the closet, the only part left standing after the tornado passed.

Although the tornado was worse than expected, meteorologists had been able to predict a severe storm that had the potential to produce tornadoes. Television and radio stations had issued alerts. Unfortunately, there is still no way to predict exactly when or where a tornado will form, nor exactly what path it will take. All of this would be useful information, particularly because the paths of destruction cut by tornadoes are relatively narrow. While hurricane damage can be predicted across hundreds of miles, a house across the street from a tornado's path may suffer no damage at all.

READING SELECTION
EXTENDING YOUR KNOWLEDGE

IMPROVING DISASTER SURVIVAL RATES

Following a natural disaster such as a tornado, most local governments review the steps in their emergency preparedness plans—or develop such a plan—that could reduce the death and destruction resulting from such a disaster. A crucial step in saving lives is to warn the population of coming storms. Most localities rely upon sirens, radio, television, telephone, and computer networks to alert local residents. In tornado-prone areas, children are taught at school what to do in response to a tornado warning. Officials and the media also encourage every family to discuss safety practices in the event of natural disasters, particularly if they live in an area prone to tornadoes or other natural disasters.

Engineers know the types of building design most likely to withstand tornadoes. They recommend the use of clips, anchor bolts, and straps to tie together the roofs, walls, and foundation of a home to prevent its collapse in a tornado. Tests of concrete walls have shown their ability to withstand flying debris, one of the most dangerous features of a tornado or hurricane. Existing buildings can be updated to provide

▶ HERE, WORKERS ARE USING SAND TO CONTAIN SPILLS OF OIL AND FUEL FROM DAMAGED AIRPLANES AFTER A TORNADO IN ARKANSAS. AN EMERGENCY PREPAREDNESS PLAN SHOULD INCLUDE MEASURES TO REDUCE SUCH RISKS.

PHOTO: FEMA/Charles Powell

WATERSPOUTS

A waterspout is a rotating column of air over a large body of water. Some waterspouts begin as tornadoes and then move over water. These are called tornado waterspouts. Other waterspouts form over a large body of warm water. These are called fair-weather waterspouts. They do not start from a thunderstorm, and they are much smaller than tornado waterspouts. Some scientists think this type of waterspout forms when sea breezes meet. Waterspouts are very common in the Florida Keys and over the Great Lakes in the summer.

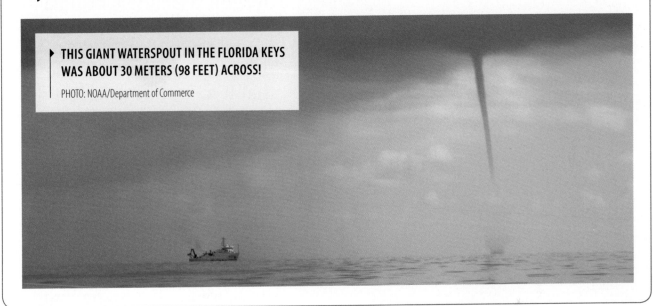

▶ THIS GIANT WATERSPOUT IN THE FLORIDA KEYS WAS ABOUT 30 METERS (98 FEET) ACROSS!

PHOTO: NOAA/Department of Commerce

more protection from moderate tornadoes, and new buildings, such as schools, police and fire stations, residences, and offices, can be built to meet stronger safety requirements for withstanding more violent storms.

An F5 tornado, however, may chew through even the strongest building. In areas where very strong tornadoes are common, many people take the precaution of building underground storm shelters in which they can safely ride out even a direct hit. ■

DISCUSSION QUESTIONS

1. Why do you think Dixie Alley, which used to be overlooked as an area of tornado danger, became prominent in the news after 2010?

2. Do you think your home is built to withstand a tornado well? Why or why not?

TEMPERATURE, PRESSURE, AND CLOUD FORMATION

▶ **HURRICANE ISABEL LASHED COMMUNITIES IN THE EASTERN UNITED STATES, UPROOTING TREES AND BRINGING THE THREAT OF FLASH FLOODS.**

PHOTO: National Oceanic and Atmospheric Administration (NOAA)

INTRODUCTION

Imagine a cloud taller than Mt. Everest—the tallest mountain in the world! Or think of a cloud so symmetrical that it looks like a perfect pinwheel in the sky. Hurricane Isabel, pictured at left, was like that—tall and symmetrical. How do clouds form? Why do they sometimes become storms? Evaporation, warm rising air, dust particles, condensation of water, and air pressure are the ingredients for making clouds. In Lessons 4 and 5, you modeled the movement of air caused by the uneven heating of the earth's surface. In this lesson, you will extend your observations to investigate how water evaporates and condenses as clouds. You will also investigate how air pressure affects cloud formation. Finally, you will relate this information to weather on the earth.

OBJECTIVES FOR THIS LESSON

Model and describe how water evaporates and condenses and how these processes play a part in cloud formation.

Model and describe the air pressure conditions under which clouds form.

Analyze weather maps, classify fronts, identify high- and low-pressure systems, and determine the weather conditions associated with each.

Collect and analyze weather data, noting any patterns involving changes in barometric pressure.

Complete and present your project on weather observations and predictions.

▶ MATERIALS FOR LESSON 6

For you

1 copy of Student Sheet 6.4: Reading Weather Maps
Your weather observations data table (from Inquiry 6.3)
1 metric ruler

For your group

1 plastic box
2 clear bottles with caps
1 flashlight
2 digital thermometers
1 beaker of hot water
1 beaker of cold water
1 ice cube
3 consecutive daily weather maps

GETTING STARTED

1 Look at the U.S. weather maps shown in Figure 6.1. Work with your group to make general observations of the maps. Discuss these questions:

A. Where on each map do you think it is cloudy? How do you think clouds form?

B. What do you think "H" and "L" on the maps represent?

C. What type of weather would you expect in an area marked with an "H"? What type of weather would you expect in an area marked with an "L"?

2 In this lesson, you will investigate some of the conditions under which clouds form. At the end of the lesson, you will look at weather maps again to relate what you observed in the lab to weather on the earth.

▶ **THREE CONSECUTIVE DAILY WEATHER MAPS**
FIGURE **6.1**

SOURCE: Forecast and graphics provided by AccuWeather.com ©2011

A. **National weather**

Shown are noon positions of weather systems and precipitation. Temperature bands are highs for the day.

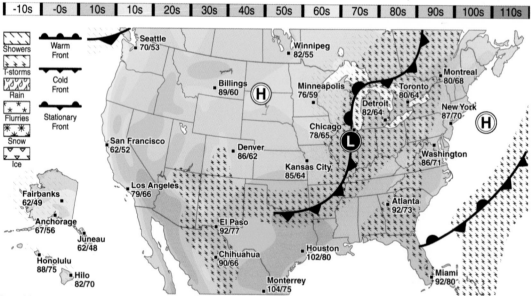

Drenching showers and locally gusty thunderstorms will stretch from the northern Great Lakes to the Ohio Valley today. Thunderstorms will gather along the southern Atlantic Seaboard, while more spotty storms will affect the Desert Southwest and southern Plains. Dry weather will hold in the coastal Northeast. Heat will build in the West.

B. National weather

Shown are noon positions of weather systems and precipitation. Temperature bands are highs for the day.

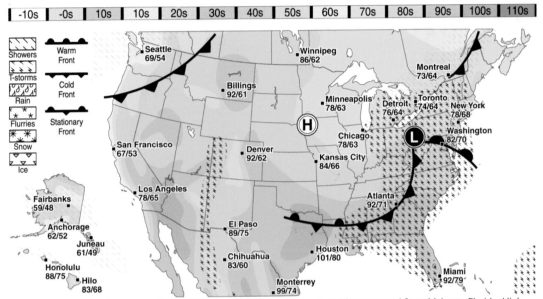

Areas of rain and thunderstorms will spread into the East today with wet weather expected from Maine to Florida. High pressure moving into the Midwest will dry the region out as afternoon storms persist across the Rockies. Much of the West will be dry, but a few showers will wet parts of western Washington.

C. National weather

Shown are noon positions of weather systems and precipitation. Temperature bands are highs for the day.

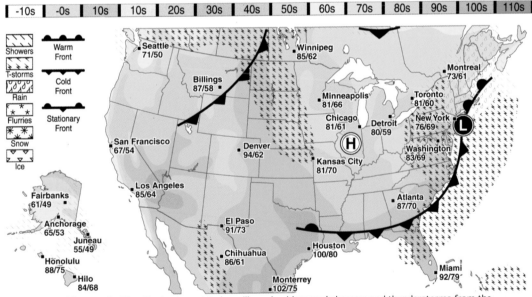

Low pressure riding up the New England coast today will spark widespread showers and thunderstorms from the Northeast through the mid-Atlantic. A few thunderstorms will rumble along the East Coast while high pressure brings dry weather to the Great Lakes and Ohio Valley. Severe storms will rumble across the northern Plains.

INQUIRY 6.1

OBSERVING EVAPORATION AND CONDENSATION

PROCEDURE

1 What do you already know about the water cycle and how clouds form? Discuss your ideas with the class. Record them in your science notebook.

2 Look at the equipment you will use in this lesson to observe changes in water due to warming or cooling. You will explore how the temperature of water affects evaporation and condensation.

3 Record the question from Step 2 in your notebook. Leave enough room to write your observations.

4 How would you use the materials to test this question? Share your ideas with the class.

5 Discuss with your teacher how you will record your predictions and observations in a table.

READING SELECTION

BUILDING YOUR UNDERSTANDING

HURRICANE FORMATION AND THE WATER CYCLE

Why do hurricanes develop over warm, tropical water near the equator? The warm water provides an almost endless supply of energy for these storms. Massive clouds form from the warm, evaporating water. Winds move the storm clouds along the hurricane's path. When the hurricane moves over land or cool northern water, it loses its energy and dies.

What process is at the heart of hurricane formation? It is the water cycle. As the illustration of the water cycle shows, during the water cycle, clouds form when warm air rises or when warm air and cold air meet. Clouds are made up of billions of tiny droplets of water or ice and dust particles. When water on the earth absorbs heat energy, it evaporates into a gas called water vapor. When air rises, it carries the water vapor with it. Air at higher altitudes within the troposphere is cool. Cool air cannot hold as much water vapor as warm air can, so some of the vapor turns into drops of liquid water. The process by which water changes from a gas to a liquid is called condensation. The process of condensation, shown in the illustration, releases heat which feeds more energy into the system and evaporates the condensed water, causing the vapor to rise even higher. The liquid water collects on dust particles and forms clouds. ∎

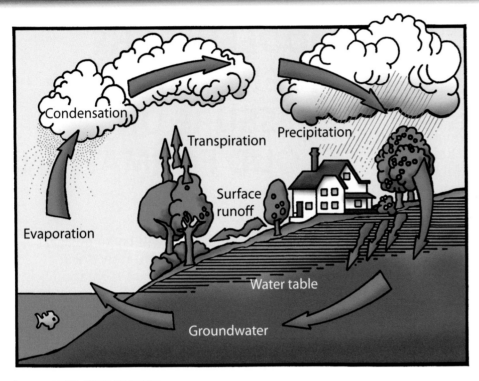

▶ **THE WATER CYCLE ON EARTH**

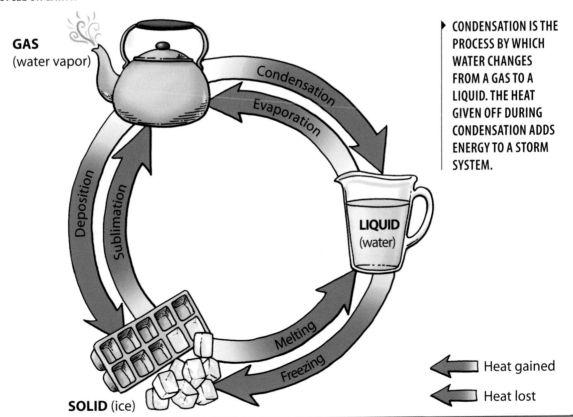

▶ CONDENSATION IS THE PROCESS BY WHICH WATER CHANGES FROM A GAS TO A LIQUID. THE HEAT GIVEN OFF DURING CONDENSATION ADDS ENERGY TO A STORM SYSTEM.

Inquiry 6.1 continued

INQUIRY 6.2

6 Collect your materials. Set them up as your group has planned. Then conduct your investigation and record your observations.

7 How does the temperature of the air affect evaporation and condensation? Try this: Rub the ice cube on the outside of each bottle. (See Figure 6.2.) What happens? Rub your hands on the outside of the bottle. What happens? Record your observations and discuss them with your group.

8 Clean up according to your teacher's instructions.

9 Read "Hurricane Formation and the Water Cycle" on pages 98-99.

▶ RUB AN ICE CUBE AND THEN THE PALMS OF YOUR HANDS ON THE OUTSIDE OF THE BOTTLE.
FIGURE **6.2**

MODELING THE EFFECTS OF AIR PRESSURE ON CLOUD FORMATION

PROCEDURE

1 In this inquiry you will try to answer the following question: How does air pressure affect cloud formation?

2 Record the question in your science notebook. 🖉

3 Discuss with the class how you will use the bottle of hot water to investigate this question on clouds. To get started, discuss these questions with your group:

A. What are the "ingredients" for cloud formation? (Think back to Inquiry 6.1 and the reading selection "Hurricane Formation and the Water Cycle.")

B. How could you create these conditions in your bottle?

C. If you want to test how air pressure affects cloud formation, how could you create high pressure in the capped bottle?

D. How could you create low pressure in the bottle?

E. How could you keep track of your predictions and observations?

4 Review Procedure Steps 5 and 6 with your teacher. Then collect your materials. Create a table to record your data and observations.

5 Your teacher will use a burning punk stick to add smoke to your bottle, as shown in Figure 6.3, keeping the lit punk inside the bottle for approximately 3 to 5 seconds. When your teacher removes the punk stick, quickly cap the bottle.

6 Before starting the investigation, swirl the water inside the bottle to reduce fog. Then shine the flashlight on the bottle, as shown in Figure 6.4, while squeezing and holding the bottle. Then, release your hands from the bottle. Record your observations.

7 When your group is finished, clean up and return your materials.

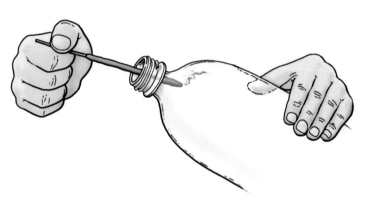

▶ YOUR TEACHER WILL USE THE PUNK STICK TO ADD SMOKE TO EACH GROUP'S BOTTLE.
FIGURE **6.3**

▶ SQUEEZE AND RELEASE THE CAPPED BOTTLE WHILE SOMEONE ELSE SHINES THE LIGHT ON IT. DO CLOUDS FORM UNDER HIGH OR LOW PRESSURE?
FIGURE **6.4**

NOTICING PATTERNS AND RELATIONSHIPS IN WEATHER DATA

PROCEDURE

1 Before weather maps existed, people made forecasts by paying attention to the weather around them. For five days, you and your group will record local weather data in your science notebook and see if you can spot any patterns or relationships that might help you forecast weather. ✍

2 You will need to draw a table in which to record your data. With your group, decide what data your table should include and how you will organize it. At a minimum, your table will need room for the following data on each day:

- high and low temperatures
- relative humidity
- total precipitation
- barometric pressure readings at 6 AM and 6 PM
- peak wind speed
- types of clouds observed

Can you think of other data you might want to include? Once you have decided, draw and title your table. Each student in your group should keep his or her own table.

3 Discuss possible sources of weather data with the class.

4 Make and record weather and cloud observations for five days, beginning today. Examine the cloud types in Figures 6.5-6.9 and in the illustration on page 110. Make sure you know what cirrus, stratus, cumulus, and cumulonimbus clouds look like. Your teacher will give the class time each day to observe and record cloud data, but you'll need to gather the rest of the data each day outside of class or for homework.

5 As you collect data, compare it with previous days' data, and discuss it with your group. Do you see any patterns emerging? For instance, does a change in one variable seem to foretell a certain kind of change in another? Do two or more variables seem to change together? If you think you see a pattern, try to make a hypothesis describing it. (For example, "When the wind starts blowing from the south, the temperature goes up.") Write it down in your science notebook. If you think you can use your ideas to make weather predictions, try it. Make sure you can explain how you made your predictions.

6 In Inquiry 6.5, you will analyze your data with your teacher's help and discuss your results and observations with the class. Now, your teacher will review the grading rubrics for the project with you. Be sure to ask questions about anything you don't understand.

▶ **CUMULUS CLOUDS OVER THE COLORADO PLATEAU**
FIGURE **6.5**
PHOTO: Sean Linehan, NOAA NOS/NGS/NOAA's National Weather Service (NWS) Collection

▶ **THIN CIRRUS CLOUDS SPREAD OUT OVER THE SKY**
FIGURE **6.6**
PHOTO: Ralph Kresge/NOAA's National Weather Service (NWS) Collection

Inquiry 6.3 continued

> **STRATUS CLOUDS**
> FIGURE **6.7**
> PHOTO: Ralph Kresge/NOAA's National Weather Service
> (NWS) Collection

> **CUMULONIMBUS CLOUDS OVER AFRICA**
> FIGURE **6.8**
> PHOTO: NASA Earth Observatory

> **STRATOCUMULUS CLOUDS OVER FLORIDA**
> FIGURE **6.9**
> PHOTO: Ralph Kresge/NOAA's National Weather Service
> (NWS) Collection

READING WEATHER MAPS

PROCEDURE

1 Working with your group, make general observations of the parts of the weather maps that you brought to class. Find the weather in your location. You can also use Figure 6.1 for this part of the inquiry.

2 Share your observations with your class and record your answers in your science notebook. 📝

3 Line up your maps in order by date. Identify one weather system (front, low-pressure system, or high-pressure system) on each map. In which direction is the system moving across the country? What do you think causes weather to move in this way? (The reading selection "Why Does the Wind Blow?" on pages 80–85 can help you answer this.)

4 Based on what you now know about weather maps, discuss the following questions with the class:

A. Do you think it would be easier to see patterns in the weather and make predictions if you could use weather maps instead of individual pieces of weather data?

B. Weather maps were first developed in the mid-1800s. Why do you think it took so long for weather maps to be invented? HINT: Where does all the data on a weather map come from? How is it collected?

5 Look over Student Sheet 6.4: Reading Weather Maps. Discuss it with your teacher. Work on it with your group. You may have to complete it for homework.

ANALYZING WEATHER DATA

PROCEDURE

1 Get out your weather observations data table and share your data with your group. Discuss any patterns you see in the data and how the data compares to your predictions.

2 Discuss with your group how to graph six variables: barometric pressure, relative humidity, peak wind speed, total precipitation, and high and low temperatures. Will you make six different graphs, or combine them? What kind of graphs will you make: line graphs or bar graphs? How will you set them up? What will the axes measure, and in what increments and units?

3 Plot your data. Do your graphs look like your group members' graphs? If not, why not?

4 Work with your group to answer these questions in your science notebook: ☞

A. Did you observe any large changes in the temperature, barometric pressure, and relative humidity over the observation period? If so, what do you think caused them?

B. Are there any relationships between your graphs? Does one variable seem to change in tandem with any others, or change ahead of any others? If you see relationships, why do you think they exist? Can any of them be used to help make weather predictions?

C. How did the clouds change over the observation period? Is there any connection between the types of clouds you saw and any of the other data you collected?

D. If you tried to predict the weather as you collected data, were your predictions accurate? What conclusions can you draw from the accuracy or inaccuracy of your predictions? What further questions did your predictions generate?

E. Did any of the data you collected seem to be unrelated to the rest? If so, why do you think that might be? Is it possible for one aspect of the weather to be unrelated to others?

5 As a group, decide what the most interesting observation was, and share it with the class. Explain why it was interesting and what it showed you about weather or weather prediction. You might discuss something strange that you saw and could not explain, why a prediction was accurate or inaccurate, or something that made you curious or gave you insight into some aspect of the weather.

REFLECTING
ON WHAT
YOU'VE DONE

1 Answer these questions and then discuss your answers with the class:

A. In Inquiry 6.1, were you able to change the amount of evaporation and condensation that occurred inside your bottles? If so, how?

B. In which bottle did you observe the most evaporation and condensation? Why do you think that happened?

C. Why did you add smoke to the bottle in Inquiry 6.2?

D. What happened to the air when you squeezed the bottle?

E. When you released the bottle, you created a low-pressure system. Describe the air inside the bottle when this happened.

F. Use your own words to describe the conditions under which clouds generally form.

2 With the class, share your responses to these questions, as outlined on Student Sheet 6.4:

A. What kind of weather is associated with a high-pressure system?

B. What kind of weather is associated with a low-pressure system?

C. What symbol represents a cold front? What symbol represents a warm front?

D. Pick one weather front on a map. What weather is associated with it?

E. In what direction does air move across the United States? Why is this information important?

3 Read "The Truth About Air" on pages 108–111.

READING SELECTION
EXTENDING YOUR KNOWLEDGE

The Truth about Air

WHY DOES A MOUNTAIN CLIMBER SOMETIMES NEED AN OXYGEN MASK? IT IS BECAUSE THE HIGHER A PERSON GOES, THE LIGHTER (LESS DENSE) THE AIR BECOMES. THAT MEANS THERE IS LESS OXYGEN FOR THE CLIMBER TO INHALE. IN THIS PICTURE, THE PARTICLES OF AIR ARE REPRESENTED BY DOTS. WHERE DO YOU THINK THE DENSITY OF AIR IS GREATER? COULD YOU MAKE ANY GENERALIZATION(S) RELATING ELEVATION ABOVE SEA LEVEL, AIR DENSITY, AND AIR PRESSURE?

AIR PRESSURE

Water pressure is the force exerted by the weight of water on a surface. Think about that for a moment. Water is pretty heavy: one kilo per liter, or about eight pounds per gallon. The weight of water just sitting high in a water tower, which may hold millions of gallons, is enough to force water through faucets many miles away.

Air pressure, likewise, is the force exerted by the weight of air on a surface. Air does have mass, and, therefore, weight. At sea level, one square inch of air exerts a pressure of about 6.7 kg (14.7 lbs). That's the weight of the entire column of the atmosphere rising above that square inch. As the elevation above sea level increases, the air becomes less dense and the "column of atmosphere" above it is shorter. This means there's less mass weighing on each square inch, and the air exerts less pressure.

TORRICELLI: INVENTOR OF THE MERCURY BAROMETER

Scientists and weathermen use barometers to detect and measure air pressure changes. It's an old device, but as useful as ever. In 1643, Evangelista Torricelli, a student of Galileo, accidentally invented the mercury barometer in his effort to create a perfect vacuum. (At the time, there was great interest in determining whether air had weight, and if so, how much. Creating a working vacuum, a vessel entirely emptied of air, that didn't leak was a major technical problem, but necessary for determining whether a completely empty vessel weighed less than one full of air.)

Torricelli created a glass tube about one meter in length that could be opened at one end while closed at the other. He removed air from the tube, submerged it in a dish of mercury, and opened the end of the tube that was in the dish. The mercury flowed into the tube, and rose nearly 76 cm (30 in) above the dish. Above the mercury was the vacuum he had hoped to achieve.

Over a period of several days, he discovered that the mercury rose and fell slightly in the tube. Torricelli correctly concluded that air outside the tube was weighing on the mercury in the dish, and that this pressure pushed the liquid up into the tube. Because there was no air in the tube, nothing resisted the mercury's rise except gravity. In other words, the height of the liquid in the tube was a measure of atmospheric pressure. The greater the atmospheric pressure, the higher the mercury moves up the tube.

Why did Torricelli use mercury instead of water, as Galileo had done? First, water could evaporate in the tube, leading to false readings. Second, water is less dense than mercury (a metal), making a cubic centimeter of water much lighter than mercury. Remember, the thing that stopped mercury from rising all the way to the top of the tube was its own weight. A less dense liquid could rise much higher. In fact, an atmospheric pressure that forced mercury 76 cm (30 in) up the tube could push water 1034 cm (34 feet) high! A barometer that tall was clearly impractical.

Torricelli's barometer was small enough to be portable, so weather observers could measure barometric pressure at different altitudes, which gave scientists of the day confirmation that air pressure did indeed vary. A unit of air pressure (no longer in use) was named the torr, after Torricelli. Today scientists measure air pressure in units called millibars—each 1/1000th (one-thousandth) of a bar, the air pressure at sea level. Thus, air pressure at sea level can be reported as 1.0 bars, or 1000 millibars (mb). That's equal to 29.63 inches of mercury.

What do you think Torricelli's mercury barometer looked like? Draw a picture of it using information from this reading selection. Then look for library or Internet resources about Torricelli and compare your drawing with the mercury barometer he invented.

READING SELECTION
EXTENDING YOUR KNOWLEDGE

AIR PRESSURE AND WEATHER

Changing air pressure is one ingredient of weather on earth. The formation of clouds, for example, is a direct result of air pressure. Low clouds, such as stratus clouds, are denser and thicker than clouds high in the sky. Clouds that are high in the sky, such as cirrus clouds, are wispy, and thinner than clouds at low altitudes.

Differences in air pressure from region to region also cause air to move, which is the definition of wind. Air flows from high-pressure regions to low-pressure regions. If you have ever walked into an air-conditioned building in the summer and felt a rush of wind against your body, you know that air pressure affects the way air moves. The difference between the high air pressure in the building and the low air pressure outside causes the air inside to rush out the door. In fact, it's being pushed out the door by the relatively high air pressure within. (The same difference in air pressure created the movement of air in your connected convection tubes.)

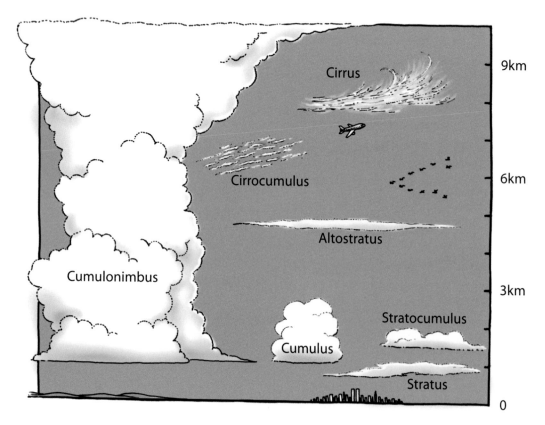

▶ AIR PRESSURE AFFECTS
THE TYPES OF CLOUDS
THAT FORM AT DIFFERENT
ALTITUDES.

Because air masses are in constant motion, air pressure in any location on earth can vary from day to day, or even from hour to hour. Since pressure differences help scientists predict how air might move, monitoring and predicting changes in air pressure is fundamental to weather forecasting. ■

DISCUSSION QUESTIONS

1. Why does a weather forecast frequently include a barometer reading?

2. If you had to design a barometer using materials available in your home or classroom, how would you do it?

HURRICANES: DESTRUCTIVE STORMS

▶ **THE STORM SURGE FROM HURRICANE ISABEL SURROUNDS HOMES ALONG THE CHESAPEAKE BAY.**

PHOTO: FEMA/Crystal Payton

INTRODUCTION

You've investigated the devastating effects of tornadoes. Even more dangerous, however, are storms that can form over warm tropical waters; hurricanes are among the earth's most destructive natural disasters. In this lesson, you will investigate the formation, behavior, and power of hurricanes.

OBJECTIVES FOR THIS LESSON

Review previous knowledge of hurricanes and their locations.

Investigate and model the factors that affect the height of a storm surge.

Learn about the behavior and effects of hurricanes.

Track the formation and movement of Hurricane Katrina.

▶ **MATERIALS FOR LESSON 7**

For you

1	copy of Student Sheet 7.1: Modeling the Storm Surge
1	copy of Student Sheet 7.2: Tracking Hurricane Katrina
1	colored pencil

For your group

1	clear tray
1	container of water, 1 L
1	metric ruler
1	flexible straw
1	skewer
1	fine-tipped permanent marker
1	syringe
1.5 lbs	modeling clay
1	laminated Weather and Climate World Map

For your class

DVD: *Hurricane Katrina: The Storm that Drowned a City*

GETTING STARTED

1 Collect and record the daily data for your weather observation project.

2 Examine your group's laminated Weather and Climate World Map from Lesson 1. Where did your group think that most hurricanes occurred? Did you omit any hurricane (or similar storm) areas?

3 Read "Hurricane Formation and Impact" on pages 115–116.

4 Look at the pictures of the hurricane on page 115. Can you identify the eye? In what direction is the hurricane spiraling? Make a drawing of the parts of the hurricane in your science notebook. 📝

▶ **SATELLITE IMAGE OF HURRICANE IRENE**

PHOTO: NASA GSFC GOES Project

HURRICANE FORMATION AND IMPACT

Hurricanes form in tropical regions with warm water and warm, moist air. When tropical waters are heated to about 27°C, not just at the surface but to a depth of about 46 meters, a hurricane can develop. Also helpful for hurricane development are a cool atmosphere, a still ocean surface, the absence of winds as the clouds climb through the atmosphere, and a distance of at least 500 km from the equator. As you can see, hurricanes can't form just anywhere, anytime. Rather, we see hurricanes forming in the same places and traveling the same paths year after year. Most hurricanes in the Atlantic have their origin in easterly winds blowing off the coast of Africa.

How do hurricanes form? As the ocean heats up, the very warm surface water evaporates and the vapor rises. When the vapor reaches higher, cooler atmosphere, it condenses to form storm clouds and raindrops. The condensation process releases a great deal of heat, which warms the cooler air above, creating a large convection cycle. The same process of water warming and evaporating, followed by vapor rising, condensing, and releasing heat continues, building clouds around a core of low pressure. This structure can produce the spinning that characterizes a hurricane. Earth's rotation helps maintain low pressure in the middle of the storm, while global east-to-west winds send it on its way across the ocean. In the Northern Hemisphere, the air circulates in a counterclockwise pattern; in the Southern Hemisphere, it circulates in a clockwise direction.

The calm, low-pressure center of a hurricane is called its "eye." Some hurricanes may have more than one calm center of low pressure, but they always have a main eye. The area on the outside of the eye is known as the "eye wall." Due to the low pressure and the narrow diameter of the storm in the eye, the winds

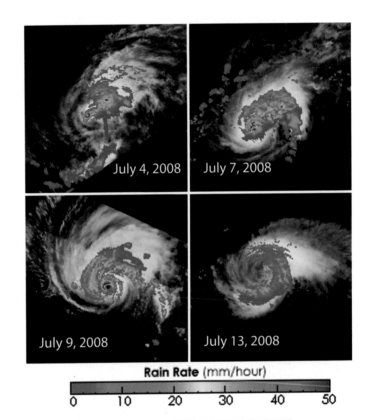

THESE IMAGES, TAKEN BY THE TROPICAL RAINFALL MEASUREMENT MISSION (TRMM) SATELLITE, SHOW THE DEVELOPMENT AND DECAY OF HURRICANE BERTHA IN THE ATLANTIC OCEAN.

PHOTOS: NASA/Images produced by Hal Pierce

there spin faster than at the outside of the storm. When the pressure in the hurricane drops, its intensity increases. If the pressure drops suddenly, the hurricane quickly becomes a very dangerous storm.

Hurricanes also tend to be very large storms, hundreds of kilometers across. That means that if they make landfall, they strike quite a bit of land at once.

READING SELECTION

BUILDING YOUR UNDERSTANDING

continued

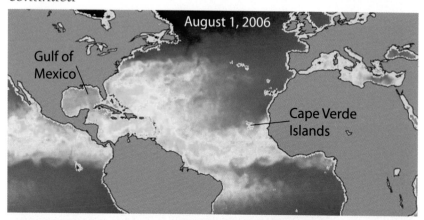

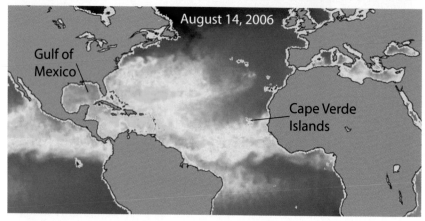

▶ THESE IMAGES, CAPTURED BY JAPAN'S ADVANCED MICROWAVE SCANNING RADIOMETER FOR EOS (AMSR-E) FLYING ON NASA'S AQUA SATELLITE, SHOW WHERE SEA SURFACE TEMPERATURES WERE HURRICANE-READY IN THE ATLANTIC OCEAN ON TWO DIFFERENT DATES IN AUGUST 2006. AREAS WHERE WATERS ARE WARM ENOUGH TO BE HURRICANE-READY ARE YELLOW OR RED. WATERS THAT ARE NOT WARM ENOUGH FOR HURRICANES TO FORM ARE SHOWN IN BLUE.

PHOTOS: Image by Jesse Allen, based on data from Remote Sensing Systems/NASA Earth Observatory

Sea Surface Temp (° C)

-2   16.5   28   35

LANDFALL AND SURGE

Hurricanes cause more damage than any other kind of storm on earth. At NOAA, scientists estimate that the heat energy thrown off every 20 minutes by a single hurricane is equivalent to the energy released by the explosion of a 10-megaton nuclear bomb.

As a hurricane strikes land (a moment called landfall), it pushes ahead of itself a hill of ocean water as high as 4.5 m or more at its peak and 160 kilometers across. The arrival of this hill of ocean water is called the storm surge. Torrential rains worsen flooding near the shore and bring floods far inland.

Storm surge and the resulting flooding cause over 90% of hurricane deaths. Hurricane Katrina, with a storm surge 7.6–8.5 meters high, was the costliest storm to strike the United States, causing nearly 2,000 deaths and approximately $80,000,000,000 worth of property losses in 2005. Homelessness after the storm was widespread, forcing many people to leave the area, some permanently. Several years later, the city of New Orleans, which took a direct hit from the storm, had a population only about two-thirds its 2005 pre-Katrina size.

One of the longest-lasting results of hurricane landfall in history hasn't been property damage, but political change. The deadliest recorded hurricane was the Bhola Cyclone, a Category 3 storm that took approximately 140,000 lives along the coast of eastern Pakistan (now Bangladesh) in 1970. (In the Pacific, hurricanes are called tropical cyclones.) The government in West Pakistan, a thousand miles away, was slow to send help to the victims, even as cholera and typhoid broke out among survivors who drank contaminated water. This callousness outraged the public and worsened the long-simmering political fights within the country. A civil war erupted, and East Pakistan broke away. It became Bangladesh, an independent country, in 1971. ■

INQUIRY 7.1

INVESTIGATING THE STORM SURGE

PROCEDURE

1 Look at the first three questions on Student Sheet 7.1: Modeling the Storm Surge. Discuss your answers with your group and record them on your student sheet.

A. What are the most important factors that determine the impact of a storm surge?

B. What type of coastal area do you think would be most vulnerable to flooding from a storm surge: a gently sloping beach or high cliffs?

C. How could you use your equipment to model the impact of the main variables affecting the storm surge?

2 Look at the drawings of coastal landforms your teacher displays. Your group will model both these shorelines, beginning with the shoreline with cliffs.

3 Have one member of your group pick up your materials. In your tray, mold the clay into a shoreline with steep cliffs as shown in the drawing and Figure 7.1(A). The landform should be about 2 cm tall and 10 cm long, and fill the width of the tray. Try to make the edge of the landform as straight as possible.

4 Use a permanent marker to mark half-centimeter increments on a skewer up to 10 centimeters, starting from the blunt end. This marked skewer will allow you to measure the height of the storm surge in hurricane conditions. Stick the blunt end of the skewer into the clay landform as close to the center edge of the shoreline as you can (see Figure 7.1). The skewer must remain standing up, oriented so you can count the centimeter marks.

▶ **SETUP FOR INQUIRY 7.1 SHOWING (A) THE STEEP CLIFFS LANDFORM AND (B) THE GENTLY SLOPING BEACH.**
FIGURE **7.1**
PHOTOS: © 2012 Carolona Biological Supply Company

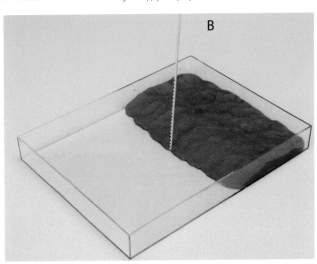

Inquiry 7.1 continued

5 Pour water into your tray until it reaches the 0.5-cm mark on the skewer. This depth represents baseline ocean conditions in your model, in the absence of a hurricane. Record "0.5 cm" in Table 1 on your student sheet as "Water Height (cm) in Non-hurricane Conditions" for "Model Landform with Cliffs." Is there any water overlapping the model shoreline? If so, use your ruler to measure the distance from the shoreline to the farthest extent of the water over the clay. Record that number (which may be zero) in the table on your student sheet as "Shoreline Flooding for Non-hurricane Conditions."

6 Model hurricane conditions. Recall that the low air pressure inside a hurricane draws ocean water up above sea level. Imagine that your tray represents the ocean at the site of a hurricane. Add 0.5 cm more water, so the water level reaches the 1.0-cm mark on the skewer.

7 Now you will model the hurricane winds that push this hill of water into a storm surge. Assign roles to each member of your group. One group member will blow gently into the shorter end of a straw to create "wind" over the water in the tray. This "windmaker" should hold the straw so that the longer end is just above the water and just a few centimeters in front of the skewer. A second group member will observe the water at the opposite end of the straw and measure the height at which the water contacts the model shoreline. A third student will use a ruler to measure how far the storm surge floods the landform. The fourth student will record the observations and measurements in Table 1 on Student Sheet 7.1.

8 Conduct three trials. The same person should create the wind for each trial.

9 Discuss your results with the class. How far inland did the water go in your modeling of non-hurricane conditions? In hurricane conditions? How high was the storm surge when it hit the coastline?

SAFETY TIP

Do not share straws. Choose one member of the group to use the straw.

10 Now you will model a gently sloping shoreline. Before you reshape the clay, use the syringe to remove as much water as possible from your tray and place it back in the water container.

11 Reshape your shoreline to match the one in the drawing your teacher displays and in Figure 7.1(B). Flatten the ocean side of your shoreline so it is less than 0.5 cm tall and the slope is as gentle as possible. Again, make the edge of the landform as straight as possible and position the skewer as close to the edge of the shoreline as you can so that it remains standing.

12 Repeat Steps 5-9 with this new landform. The same windmaker should produce the wind for this landform. Record all results on Student Sheet 7.1 in the portion of the table for "Model Landform with Gentle Slope."

13 Have the windmaker experiment with creating different wind strengths over the sloping landform. Do your results change with different wind strengths? Discuss your findings with the class.

14 Complete questions D-I on Student Sheet 7.1. Then, discuss your answers with the class.

D. In your model, how did hurricane conditions affect water height? What does this change in water height represent?

E. How did hurricane conditions affect shoreline flooding?

F. Which type of coastal area did the model show was most vulnerable to flooding from a storm surge: gently sloping or with high cliffs?

G. How do you think your model could be improved?

H. What types of structures might you add to a shoreline to reduce the impact of a storm surge?

15 Clean up your work area.

16 Read and discuss "2005—The Year of Hurricane Firsts" on pages 122-125.

INQUIRY **7.2**

TRACKING HURRICANE KATRINA

PROCEDURE

1 You will use Table 1 on Student Sheet 7.2: Tracking Hurricane Katrina and a colored pencil to track Hurricane Katrina on the hurricane tracking chart. Review with your group the meaning of latitude and longitude.

2 For each day and time included on Table 1, use a colored pencil to plot on the map the intersection of the latitude and longitude provided. Then connect the dots as described in Step 1 on Student Sheet 7.2.

3 Record your answers to the following questions under Step 2 on Student Sheet 7.2 and be prepared to discuss your answers with the class.

A. Where did the tropical storm that eventually turned into Katrina begin? Is this where tropical storms are usually born?

B. At what point (longitude and latitude) did the tropical storm become a hurricane?

C. In what direction did the storm move?

D. Look back to the reading selection "Why Does the Wind Blow?" in Lesson 5. What do you think caused Hurricane Katrina to move along this path?

E. Where did Hurricane Katrina lose its energy and turn back into a tropical storm? Why do you think it happened in that location?

F. If you had been working at the National Hurricane Center when Hurricane Katrina struck, which cities or areas would you have evacuated? What day would you have requested the evacuation? Why?

G. What happened to the wind speed and the barometric pressure over the eight-day period?

H. What data is missing that might explain the extensive damage caused on the Gulf Coast and New Orleans by Hurricane Katrina?

4 Read "The Rise and Fall of Katrina" on pages 126–131. Examine the forecasts made as the hurricane took shape and crossed into the Gulf. Then answer these questions in your science notebook. Be prepared to share them with the class: 🖉

• How many days did it take meteorologists to make an accurate prediction of where Katrina would make landfall along the Gulf Coast?

• Why did it take so long?

• Why didn't meteorologists know right away how strong the hurricane would be?

5 View the DVD on Hurricane Katrina, shown by your teacher. As you watch, take notes on the information that seems most striking and surprising to you. Be prepared to share your notes with your group and the class.

REFLECTING
ON WHAT
YOU'VE DONE

1 Answer these questions in your science notebook and then share your answers with the class.

A. In Inquiry 7.1, what were the independent and dependent variables? What was the relationship between the height of the surge and the other variables?

B. What plans to you think a family should make before a disaster such as a hurricane or tornado strikes?

C. What plans do you think a local government (city, town, county) should make in advance of a disaster?

2005—
THE YEAR OF HURRICANE
FIRSTS

The Atlantic hurricane season of 2005 was record-setting. Records were set for:

- the number of tropical storms (28)
- the number of hurricanes (15)
- the number of major hurricanes (7)
- the number of Category 5 hurricanes (windspeeds over 250 kph [155 mph]) (4)
- the number of tropical storms developing before August 1st
- the highest dollar value of damage done in the Western Hemisphere ($80,000,000,000 from Hurricane Katrina alone)

FROM ARLENE TO ZETA

There were 26 named storms in 2005, from Arlene to Zeta. Among them were many unusually strong, destructive, and long-lived storms.

- Hurricane Dennis formed in the Gulf of Mexico, rather than off the western coast of Africa, where most tropical storms begin. It crossed the south Caribbean island of Grenada and struck Cuba and Haiti.

- Hurricane Emily, the first Category 5 storm, broke a record for hurricane intensity in July. It caused extensive damage across the Yucatán Peninsula in Mexico, and killed at least 14 people.

- Hurricane Katrina formed in the Bahamas, unusually close to the United States, and within six days became the deadliest hurricane to hit the U.S. since 1928 and the most destructive on record.

- Before Hurricane Maria petered out, its remnants struck Norway, causing landslides and one fatality. This is very unusual, since most hurricanes are reduced to tropical storms and then fall apart long before they finish their northeastward hook across the Northern Atlantic.

▶ **FLORIDA HOME DESTROYED BY FLOODING FROM HURRICANE DENNIS**

PHOTO: FEMA/Andrea Booher

▶ **VOLUNTEERS WORK TO REPAIR A ROOF DAMAGED BY HURRICANE DENNIS**

PHOTO: FEMA/Leif Skoogfors

• Hurricane Ophelia swept up the East Coast of the U.S. and hit Nova Scotia. Again, most hurricanes subside long before they get to the cold waters of Canada.

• Hurricane Rita, another Category 5 storm, destroyed offshore oil platforms in the Gulf of Mexico and spawned over 100 tornadoes as it moved through Texas and Louisiana.

• Hurricane Wilma had the lowest barometric pressure ever measured (882 mb), and developed into the strongest hurricane ever recorded in the Atlantic. It struck the Mexican state of Quintana Roo as a Category 4 storm, killing 23 people.

TEMPERATURE RISE IN THE CARIBBEAN SEA

Why were there so many destructive storms in one year? Meteorologists say it was because of the wind patterns and the unusually warm temperatures in the Caribbean Sea in the spring and summer of 2005. The sea temperature was about 2°C warmer than usual, and there was relatively low wind shear. (Wind shear is the change in wind speed and direction that occurs over a very short distance. The winds where hurricanes form tend to be smooth, not choppy.) Both conditions favored the formation of tropical storms and hurricanes.

KATRINA: DEADLY AND COSTLY

Of all the hurricanes in 2005, the most dramatic was Hurricane Katrina. It ranks with storms such as the Galveston, Texas, hurricane of 1905 that demolished a city and claimed 8000 lives, and with the Florida Keys hurricane of 1935 that destroyed a railroad and trains carrying passengers in the sparsely settled islands. Katrina was not only devastating to the city of New Orleans, it also destroyed small towns and cities along the Gulf Coast from Texas to the Florida panhandle. In addition to huge property losses, Hurricane Katrina caused the deaths of an estimated 1800 people during the storm and its subsequent flooding.

READING SELECTION
EXTENDING YOUR KNOWLEDGE

While hundreds of thousands of residents left the predicted landfall area before the storm arrived, tens of thousands were without transportation and could not leave; others simply decided to ride out the storm, and became trapped when it arrived. Relief shelters were woefully inadequate, even dangerous. These problems sparked intense debates about the disaster planning and the ways in which governments should react in a catastrophe. Citizens, government workers, and meteorologists continue to debate the most effective ways to prevent the damage caused by hurricanes. One of the results has been the installation of improved pumping systems for New Orleans, which were used in 2011 during the landfall of Tropical Storm Lee. ∎

▶ BEFORE-AND-AFTER PHOTOS SHOW THE DAMAGE WROUGHT BY HURRICANE KATRINA IN BILOXI, MISSISSIPPI.

PHOTO: FEMA/Mark Wolfe

▶ CLEANING UP DEBRIS IN ST. BERNARD PARISH, LOUISIANA, AFTER HURRICANE KATRINA

PHOTO: FEMA/Win Henderson

▶ SCHOOLS HELPED PROVIDE TRAILERS AS TEMPORARY HOUSING FOR THEIR TEACHERS WHOSE HOMES WERE DESTROYED BY KATRINA.

PHOTO: FEMA/Marvin Nauman

▶ WORKERS ARE RECONSTRUCTING A RAILROAD BRIDGE IN COASTAL MISSISSIPPI THAT WAS DESTROYED BY KATRINA.

PHOTO: FEMA/Mark Wolfe

DISCUSSION QUESTIONS

1. Why were there so many hurricanes and tropical storms in 2005?

2. What factors combine to make a hurricane destructive to a particular geographic area?

THE RISE AND FALL OF KATRINA

In 2001, the magazine *Scientific American* described New Orleans as a "disaster waiting to happen." The magazine said a direct hit from a hurricane was "inevitable," that thousands could die, and that the city might be submerged under 6 meters (20 feet) of water. Why the dire warnings? Because New Orleans was protected from the waters of the Mississippi River and Lake Ponchartrain only by a system of levees, or dikes. Engineers were concerned the levees were inadequate for the storm surge from a major hurricane.

The prediction was realized in full when Hurricane Katrina hit New Orleans on August 29, 2005. So many people lost their homes that today the city has two-thirds the population it had before the storm. Neither the city nor the nation had taken the threat seriously enough. But it's hard to take hurricanes seriously before they exist, and once they do, they don't give much warning.

Katrina's journey to New Orleans took six days, beginning on August 23. Unlike many hurricanes, it was not born off the coast of Africa; rather, it began as the remnant of an earlier tropical depression that had fallen apart. Called Tropical Depression 12 (TD 12), it hit the radar close to North America, 280 kilometers (175 miles) northeast of Nassau in the Bahamas.

By the next day, TD 12 had reached tropical storm status, and had a name. Tropical Storm Katrina was moving slowly over the Bahamas, and forecasters expected it would hit Florida. Models differed on whether it would move up the Florida coast or cross into the Gulf.

▶ **SATELLITE IMAGE OF TROPICAL DEPRESSION 12 THREE DAYS BEFORE IT BECAME HURRICANE KATRINA**

PHOTO: National Oceanic and Atmospheric Administration (NOAA)

▶ **HURRICANE-STRENGTH KATRINA ABOUT TO CROSS OVER FLORIDA**

PHOTO: NASA image by Jeff Schmaltz, MODIS Land Rapid Response Team at NASA GSFC

Forecasters were also unsure how strong the storm might be. They figured it might hit the coast at moderate intensity and speed:

AT LEAST STEADY INTENSIFICATION OF A NORMAL RATE OF 10 KT PER 12 HOURS UNTIL LANDFALL OCCURS SEEMS JUSTIFIED.

("KT" means knots; a knot is a nautical unit of speed, a little more than one mile per hour.) Forecasters thought it might be a moderate storm because the air around Katrina happened to be dry, which was interfering with convection in the storm's core. If that air became humid, the storm could look quite different, wrote the forecasters.

This was only four days before disaster would strike New Orleans, and no one knew what was to come.

Katrina did intensify into a hurricane the afternoon of August 25, and made landfall on the east coast of Florida at 6:30 PM. For the first time, meteorological services in the U.S. and Canada saw Katrina as a serious storm. They recorded a low pressure of 985 mb in the eye of the storm; the lower the pressure in the eye, the stronger the winds tend to be on the outside. The National Weather Service (NWS) issued this report:

KATRINA IS EXPECTED TO GRADUALLY STRENGTHEN ONCE IN THE GULF OF MEXICO. THE GFDL [Geophysical Fluid Dynamics Laboratory] AND SHIPS [Statistical Hurricane Intensity Prediction Schemes] MODELS BRING KATRINA TO A MAJOR HURRICANE ... THE ECMWF [European Center for Medium-Range Weather Forecasts] MODEL DROPS THE PRESSURE OF KATRINA IN THE GULF OF MEXICO TO 961 MB ... ALL INDICATIONS ARE THAT KATRINA WILL BE A DANGEROUS HURRICANE IN THE NORTHEASTERN GULF OF MEXICO IN ABOUT 3 DAYS.

New Orleans is in the north-central Gulf. Twelve hours after this report, on the morning of the August 26, the NWS gave its first indication that New Orleans might be in the path of the storm, although forecasting models didn't agree.

THE TIMING OF THE EROSION OF THE RIDGE AND AN INDUCED NORTHWARD MOTION OF KATRINA IS THE MAIN DIFFERENCE BETWEEN THE MODELS ... THE NOGAPS [Navy Operational Global Atmospheric Prediction Systems] AND GFDN [Geophysical Fluid Dynamics Navy] MODELS HAVE MADE A LARGE JUMP TO THE WEST OVER LOUISIANA ... WHEREAS THE MAJORITY OF THE NHC [National Hurricane Center] MODELS TAKE KATRINA INLAND OVER THE NORTHEAST GULF COAST.

▶ **PHOTO OF HURRICANE KATRINA TAKEN FROM NOAA HURRICANE HUNTER AIRCRAFT**

PHOTO: National Oceanic and Atmospheric Administration (NOAA)

▶ **HURRICANE KATRINA MOVING OVER THE GULF OF MEXICO**

PHOTO: National Oceanic and Atmospheric Administration (NOAA)

▶ **HURRICANE KATRINA SHORTLY AFTER LANDFALL IN LOUISIANA**

PHOTO: National Oceanic and Atmospheric Administration (NOAA)

Early Saturday morning, August 27th, the NWS put out the alert that Katrina, with winds of 185 kph (115 mph), was a major hurricane with a visible eye and minimum pressure of 945 mb. At 10:00 AM, the NWS notified those on the Gulf Coast of a hurricane watch. For the first time, New Orleans was named specifically. Pressure at the eye was still dropping. The National Oceanic and Atmospheric Administration's (NOAA) and the Air Force's planes were measuring the storm with microwave radiometry and barometry, monitoring wind speeds, air pressure, storm structure, and the size of the eye.

A HURRICANE WATCH IS IN EFFECT FOR THE SOUTHEASTERN COAST OF LOUISIANA EAST OF MORGAN CITY TO THE MOUTH OF THE PEARL RIVER ... INCLUDING METROPOLITAN NEW ORLEANS AND LAKE PONCHARTRAIN. A HURRICANE WATCH MEANS THAT HURRICANE CONDITIONS ARE POSSIBLE WITHIN THE WATCH AREA ... GENERALLY WITHIN 36 HOURS.

The National Hurricane Center's director spoke to the *Times-Picayune* in New Orleans. "This is really scary," he said. "This is not a test, as your governor said earlier today. This is the real thing." The *Times-Picayune* went to Louisiana State University's meteorologists, looking for something more specific. What they got was a map of projected flooding in New Orleans, and the headline shouted: NO MARGIN FOR ERROR. A frightening scenario, they said: a wall of water pushing up against the levees.

The odds of the eye passing near New Orleans were up to 19 percent ... then 21 percent ... then 26 percent. Midnight hadn't yet passed. The NWS issued another warning:

A HURRICANE WARNING HAS BEEN ISSUED FOR THE NORTH CENTRAL GULF ... INCLUDING THE CITY OF NEW ORLEANS AND LAKE PONTCHARTRAIN ... HURRICANE CONDITIONS ARE EXPECTED WITHIN THE WARNING AREA WITHIN THE NEXT 24 HOURS. PREPARATIONS TO PROTECT LIFE AND PROPERTY SHOULD BE RUSHED TO COMPLETION ... COASTAL STORM SURGE FLOODING OF 4.6 TO 6.1 METERS (15 TO 20 FEET) ABOVE NORMAL TIDE LEVELS ... LOCALLY AS HIGH AS 4.6 METERS (25 FEET) ALONG WITH LARGE AND DANGEROUS BATTERING WAVES ... CAN BE EXPECTED NEAR AND TO THE EAST OF WHERE THE CENTER MAKES LANDFALL.

Sunday came. The storm strengthened in the Gulf and became a Category 4 hurricane, then a Category 5. In New Orleans, people who could leave were packing up and hitting the road. The NWS continued, intently, to monitor the hurricane's structure and the wind's strength, as well as air masses around Katrina that might push it this way or that. The probability that the storm's eye would near New Orleans was now 35 percent.

WE MUST CONTINUE TO STRESS THAT THE HURRICANE IS NOT JUST A POINT ON THE MAP ... BECAUSE DESTRUCTIVE WINDS ... TORRENTIAL RAINS ... STORM SURGE ... AND DANGEROUS WAVES EXTEND WELL AWAY FROM THE EYE. IT IS IMPOSSIBLE TO SPECIFY WHICH COUNTY OR PARISH WILL EXPERIENCE THE WORST WEATHER.

READING SELECTION

EXTENDING YOUR KNOWLEDGE

At 4:00 PM on Sunday there was a 47 percent likelihood of the eye's nearing New Orleans. At 10:00 PM this likelihood had increased to 59 percent. The question in New Orleans was, Will the hurricane come straight through our city? Upstate, people wanted to know, Where will it go afterwards? The meteorologists wrote:

KATRINA IS EXPECTED TO GRADUALLY TURN NORTHWARD INTO A BREAK IN THE SUBTROPICAL RIDGE ASSOCIATED WITH A LARGE MID-LATITUDE CYCLONE NEAR THE GREAT LAKES. MODEL GUIDANCE REMAINS TIGHTLY CLUSTERED ... WHILE THERE IS GREAT SIGNIFICANCE FOR THE CITY OF NEW ORLEANS IN THE DETAILS OF THE TRACK ... TRACK ANOMALIES OF 48–80 KILOMETERS (30–50 MILES) ARE STILL POSSIBLE EVEN 12–18 HOURS OUT.

We just can't tell you for sure, they were saying. It could hit you directly, or it could go wide. Best to get out. At 6:00 AM on Monday, the NWS told the people of New Orleans their time was up:

EXTREMELY DANGEROUS ... HURRICANE KATRINA PREPARING TO MOVE ONSHORE NEAR SOUTHERN PLAQUEMINES PARISH, LOUISIANA, ... HURRICANE-FORCE WIND GUSTS OCCURRING OVER MOST OF SOUTHEASTERN LOUISIANA ... PREPARATIONS TO PROTECT LIFE AND PROPERTY SHOULD HAVE BEEN COMPLETED.

Katrina made landfall a few hours later, and the NWS continued reporting the storm's force and direction as it slowly moved north and then north-northeast over Mississippi, before turning towards Tennessee, spawning 13 tornadoes along the way. Winds downed trees, peeled off roofs, killed livestock, and knocked down power lines and weaker buildings.

By 4:00 PM Katrina had been downgraded to a Category 1 hurricane, though its force was felt right up to Minnesota. By 10:00 PM it was only a tropical storm: still dangerous, but not catastrophic. However, Lake Ponchartrain and the storm surge had spilled over the levees, turning New Orleans into part of the Gulf of Mexico. About 2000 people would die; a major American city would be badly damaged, perhaps permanently.

Both the limits of forecasting and its triumphs are part of this story. NWS meteorologists could not be sure that a major hurricane would hit the north-central Gulf shore until two days before landfall. They could not pinpoint exactly where the hurricane would go. But they were able to put out strong enough warnings, and nearly half a million New Orleans residents left or sought shelter from the storm. So did hundreds of thousands more living along the Gulf coast.

With more knowledge about how the atmosphere works and interacts with surface waters, perhaps scientists will be able to forecast major hurricanes earlier and with more precision. It may be possible to forecast with great confidence which areas will be hit by major hurricanes. In addition to improving prediction models, however, it is important that cities have emergency preparedness plans. Remember that in 2001, scientists were predicting a major hurricane would hit New Orleans, yet the city did not move back from the shore or strengthen its levees. Nor was the nation well enough prepared for disaster in New Orleans. ∎

▶ NEW ORLEANS NEIGHBORHOOD FLOODED BY STORM SURGE THAT OVERTOPPED THE LEVEES

PHOTO: National Oceanic and Atmospheric Administration (NOAA)

DISCUSSION QUESTIONS

1. Why did the *Scientific American* writer believe New Orleans would be badly impacted by Hurricane Katrina? Name as many reasons as you can.

2. What sorts of data were collected that helped to forecast and track Hurricane Katrina?

EARTH: AN OCEAN PLANET

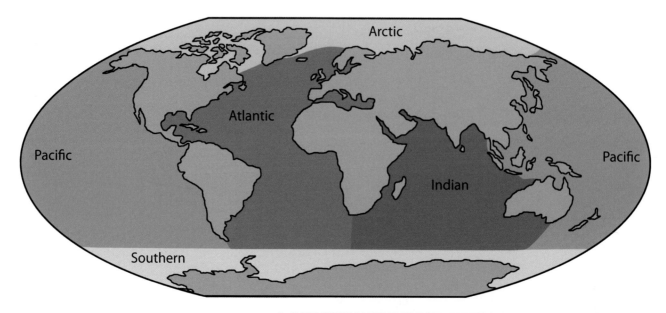

▶ **MOST OF THE EARTH IS COVERED BY OCEAN, WHICH CAN BE CATEGORIZED INTO FIVE MAIN BASINS (GEOLOGIC DEPRESSIONS THAT DIP BELOW SEA LEVEL).**

PHOTO: Based on graphic by NOAA

INTRODUCTION

Water appears, from outer space, as the dominant feature of our planet. Most of the earth's water is in the oceans. Scientists have found the oldest forms of life in the oceans and believe that life originated there. Ancient marine life was photosynthetic, and—like plants today—these organisms gave off oxygen as waste, producing an oxygen-rich atmosphere. The oxygen they released came from water: H_2O. Life took advantage of this new atmosphere: oxygen-breathing creatures evolved and thrived. In other words, the oceans are crucial to both marine life and life on land.

In the next two lessons you will investigate the oceans' depths, currents, and diversity of life, their contribution to earth's weather and climate, and the danger-fraught history of sea exploration.

OBJECTIVES FOR THIS LESSON

Locate various zones of the ocean and the features of each.

Learn to read contour maps.

Use sets of sounding marks to draw contour lines and create bathymetric maps.

Take soundings and develop a bathymetric map.

▶ MATERIALS FOR LESSON 8

For you

1	copy of Student Sheet 8.1a: Drawing Bathymetric Maps I
1	copy of Student Sheet 8.1b: Drawing Bathymetric Maps II
1	copy of Student Sheet 8.1c: Thinking About Bathymetric Maps

For your group

1	small white bucket, numbered
2	sheets of blank paper
1	wooden skewer
1	metric ruler
1.5 lbs	modeling clay
	Masking tape

GETTING STARTED

1 What do you think the ocean floor looks like? How deep is the deepest part of the ocean? Answer these questions in your science notebook. 🖉

2 Look at the Profile of the Ocean. A map that shows the depth of objects under water is called a bathymetric map, from the Greek word "bathos" for depth. A topographic map, in contrast, shows heights above sea level. Using the illustration, record your answers to these questions:

A. What are the names of the different zones of the ocean?

B. The term "pelagic" refers to the open sea. Can you guess the meaning of the prefixes "epi-," "meso-," and abysso-"?

C. How deep does the epipelagic zone of the ocean extend? The hadal zone?

D. In what ways does the ocean floor resemble dry land?

E. What would you expect the temperature to be in the hadal zone compared with that of the epipelagic zone? Why?

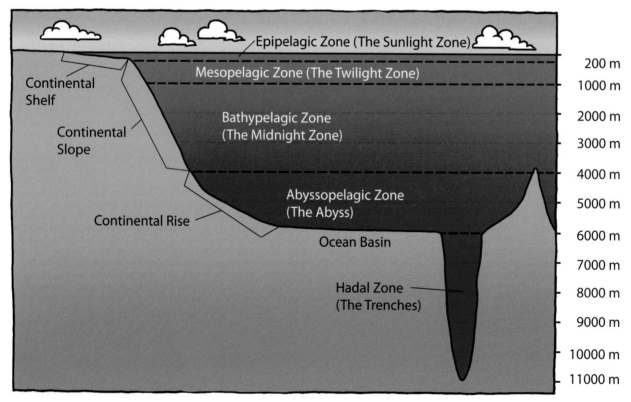

▶ **PROFILE OF THE OCEAN**

PHOTO: Based on graphic by NOAA

3 Your teacher will show you some examples of contour maps and explain how they are read. If you have any questions, ask them now. Being able to interpret a contour map is essential for what you will do in the rest of the lesson. In this lesson, you will learn to draw bathymetric maps and map a model seafloor, getting a taste of the challenge early ocean explorers faced as they tried to understand what lay beneath the waves.

PRACTICING DRAWING BATHYMETRIC MAPS

PROCEDURE

1 Listen to your teacher's explanation of bathymetric maps and how to interpret them. With the class, answer the following questions:

A. If "bathy" means deep, what does "bathymetric" mean?

B. On a bathymetric map, if a contour line is marked as 300 m, it means 300 m below sea level. What would a 400-m contour line show? Would it be higher or deeper than the 300-m line?

C. Which parts of the Ocean Profile on page 134 would be represented by contour lines that are close together on a bathymetric map? Which parts would be represented by contour lines farther apart?

D. Note that the numbers on the right side of the diagram increase as the ocean gets deeper. Which contour lines would have higher numbers, those in the hadal zone or those showing the continental slope?

Inquiry 8.1 continued

2 On Student Sheet 8.1a: Drawing Bathymetric Maps I, draw contour lines by connecting each group of depth sounding marks. Connect all the dots marked "200," then all the dots marked "300," and so on until you have drawn the entire map. If you are not sure how to connect some of the dots, ask someone in your group for help.

3 As a group, find the steepest area shown on the map. Circle it and write "steep" above the circle. Find the flattest area shown on the map. Circle it and write "flat" above the circle.

4 Compare your map with other students' maps, and then to the map your teacher displays. If your map looks different, try to figure out why. Ask your teacher for help if you are not sure.

5 You will now draw a more challenging bathymetric map. As a group, take a close look at the depth markings on Student Sheet 8.1b: Drawing Bathymetric Maps II. Notice that all of the depths marked are unique.

6 Decide with your group how you will turn this set of points into a bathymetric map. Your teacher will discuss some options with you. You will need to use contour lines that are no more than 100 meters in depth apart. You may use regular intervals—100 m, 200 m, 300 m, and so on—and each contour line must represent a single depth. You will have to decide how to draw the contour lines. Will you round the numbers and connect them? Will you draw the lines between depth soundings, instead of connecting them?

7 When you have decided on your method for drawing the map, record your plan in your science notebook. 🖉

8 Draw your bathymetric map using the method you have decided upon.

9 As in Step 3, circle and mark steepest and flattest areas on the map.

10 Compare your map with the maps created by your classmates and the maps displayed by your teacher. Discuss which methods produced the most accurate maps.

11 With your group, evaluate your method for drawing your contour lines and record your evaluation below your plan.

- How similar is your map to the one displayed by your teacher?

- Did your method work well?

- How could it have been improved?

12 If the data points you used represented a real seafloor, how could you figure out how accurate your map was? How would the scientists on the HMS *Challenger* have been able to check the accuracy of their maps in 1874? With the class, discuss how the mapping those early scientists did compares to the mapping we might do today.

INQUIRY **8.2**

INVESTIGATING THE OCEAN FLOOR

PROCEDURE

1. Look back at the diagram of the ocean profile. With your group, answer this question in your science notebook 📝:

   - Which seafloor features will you model in your container? Choose at least two.

2. Have one member of your group pick up your materials.

3. As a group, work together to build a seafloor out of clay for the inside of your bucket (see Figure 8.1). Incorporate the features you selected and exaggerate them as much as possible (taller mountains, deeper canyons). You may cover some or all of the bottom of the container.

4. Cover your bucket with paper and tape the paper in place as a lid. Try to make the covering smooth and wrinkle-free. Do not tear the paper or cover the number on the bucket. This paper will represent the surface of the ocean (or "sea level") in your model.

5. Follow your teacher's instructions for receiving the seafloor you will map. When you have it, trace the bottom of the bucket onto a blank sheet of paper. Inside the traced circle you will mark your depth soundings to create a bathymetric map. Write the bucket's number next to the traced circle.

6. Do not start taking soundings yet. Listen to your teacher's directions for how to take soundings of your seafloor. It is important to first punch a hole with the sharp end of the skewer, but then use the blunt end to feel the depth. Remember that the paper represents sea level. The points whose depth you measure will be the basis of your group's bathymetric map.

▶ **EXAMPLE OF A MODEL SEAFLOOR**
FIGURE **8.1**

PHOTO: ©2011 Carolina Biological Supply Company

7 Think about how you will measure and keep track of where you are measuring. Within your group, work out a strategy for orientation and for recording soundings on your map. Think about what landmarks you could draw on your seafloor model and your map to help you know where you took soundings.

8 Take as many soundings as you can in the time allotted. While one student places the skewer in the container and determines the depth, another student should record the sounding on your map. Be sure to mark both the location of the sounding and its depth. Other group members should help make sure that the group is making and recording accurate measurements. Take turns with these tasks, switching every few soundings.

9 When your teacher tells you to stop taking soundings, discuss as a group how you will draw the map's contour lines. When you have decided, draw the lines on your map.

10 Share your completed seafloor map with the class. Compare the maps with the seafloors. Can you tell which map represents each seafloor? What techniques created the most accurate maps? Why do you think early seafloor maps were so much less accurate and detailed than the maps we have today?

REFLECTING
ON WHAT
YOU'VE DONE

1 With the class, discuss these questions:

A. Have your ideas about the ocean changed during this lesson? If so, how do you see the ocean differently now?

B. What problems could you imagine for a seafloor-mapping expedition carried out by individual sailing ships?

C. Why is a good seafloor map useful?

D. Globes and maps have longitude and latitude lines. How is knowing longitude and latitude helpful in mapping the ocean floor?

E. What new things have you learned about data collection from doing your seafloor mapping? Is more data always useful? Can you know before you collect it whether or not it will be useful?

F. Why do scientists spend a good deal of time planning their data collection—how much, which kinds, what for—before they actually collect it?

Throwing a Grid over the Waters:
Early Oceanography

Humans have been studying the oceans for tens of thousands of years. For most of that time, as far as we know, it was desperately practical work: How do we get around the sea alive? Fishing, diving, and migration from island to island all required knowledge of a vast, dangerous sea.

As nautical skills, shipbuilding, and mapmaking improved, as trade between continents grew, and as nations went to war at sea, new questions arose: Does this ocean go on forever? How deep is this water? How big a ship is it safe to sail? How can we go faster?

Here, we'll take a tour of the early days of oceanography, the study of the oceans, which focused on maps and navigation.

EARLY SEAFARING

Many ancient societies made their livings from the sea. Two impressive seagoing nations were Polynesia and Greece.

POLYNESIA

Look at a map, and you'll see that the South Pacific islands making up Polynesia are far from any large land mass. So how did people get there? The ancestors of Polynesians came from China about 8,000 years ago. They made their way from island to island in small boats, so they had already developed some seafaring skills. By about 3,000 years ago, they had become accomplished navigators, but they didn't use anything we would recognize as navigational instruments. Instead, they observed the waves' patterns, directions, and colors. They tracked winds, clouds, stars, weather, and birds to find their way.

How do we know Polynesians were such fine navigators? Although they had no written language, they had a storytelling culture, and stories of their voyages to Hawaii survived. Given that Hawaii is about 2500 miles from the Polynesian island of Tahiti, the Polynesian stories were thought to be lore. Then in 2007

▶ **POLYNESIA**

PHOTO: Dr. Anthony R. Picciolo, NOAA NODC/Department of Commerce

researchers found a stone adze (a cutting tool) on an island on the Polynesian route to Hawaii. The adze was made from rock that could have come only from Hawaii, providing evidence that the Polynesians really had traveled back and forth to Hawaii.

GREECE

When the Greeks sailed beyond the Straits of Gibraltar, they found an enormous sea—the Atlantic Ocean—and found themselves caught up in a strong, broad current that ran from north to south. The Greeks called this current "okeano,"

their word for "river." What they had actually found was the Gulf Stream, but the name stuck, and is the root of the English word, "ocean."

The Greeks were mathematicians, mapmakers, and astronomers as well as sailors, and Greek mariners had sailed as far as the Arctic Circle by around 300 B.C., mapping Europe's coast, islands, and bodies of water as they went. Their mathematics had suggested, based on geometric calculations, that the earth was round, and they were able to approximate its size fairly accurately.

READING SELECTION
EXTENDING YOUR KNOWLEDGE

▶ **A VIKING SHIP IN THE ICE NEAR GREENLAND**

PHOTO: Library of Congress, Prints & Photographs Division, LC-USZ62-3032

▶ **REPLICA OF A CHINESE "JUNK" (SHIP)**

PHOTO: Captain Hubert A. Paton, C&GS/NOAA/Department of Commerce

TO THE ENDS OF THE EARTH

As the technology for building large seagoing ships spread in the Middle Ages, nations began making longer, more ambitious voyages. Several kingdoms stood out in spending enormous sums on ships that roamed overseas to gain power and goods.

THE VIKINGS

The Norse, or Vikings, were Scandinavian raiders, traders, and fighters who sailed from approximately 850 to 1100 A.D. Famous for their swift, maneuverable, shallow-hulled longboats, they built colonies in what are now Greenland, Iceland, and Newfoundland, Canada. Their trade routes took them as far as Baghdad. We know they navigated by the stars, and that they had crude shadow-casting instruments that could tell them how far north they were.

CHINA

In the early 1400s, the Chinese sent hundreds of technically advanced ships, with nearly 30,000 men, to explore the Indian Ocean to the west, and Indonesia to the east. The western voyages went as far as the Cape of Good Hope in South Africa, then beyond into the Atlantic Ocean. Under the leadership of Admiral Zheng He, these ships were on a mission to explore the world and show off Chinese civilization and wealth. It is thought that more than 20,000 crew sailed this armada of ships. It was an expensive mission, and it didn't last long, but in that time China showed itself to be a rich, powerful empire.

CHRISTOPHER COLUMBUS (BEARDED) EXPLAINING HIS DISCOVERIES TO THE KING AND QUEEN.

PHOTO: Library of Congress, Prints & Photographs Division, LC-USZ62-3035

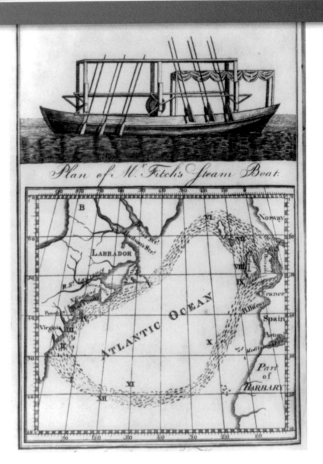

WHILE LATITUDE AND LONGITUDE LINES WERE INCLUDED ON A MID-1700S MAP USED BY FISHERMEN TO TRACK THE RUNS OF HERRING-FISH AROUND THE ATLANTIC, THEY WERE NOT LIKELY VERY ACCURATE.

PHOTO: Library of Congress, Prints & Photographs Division, LC-USZ62-31148

SPAIN AND PORTUGAL

Around 1400, European explorers began to look for faster trade routes through the ocean to Asia and Africa because their trade routes by land went over high mountains and other rugged terrain. A single trip to China, for instance, could take years.

In Spain, King Ferdinand and Queen Isabella sponsored an explorer who sought a westward shortcut to China. He, like many others, believed that the ocean must go around the globe, and stop on China's shores. In 1492, Christopher Columbus set off in a convoy of Spanish ships. His crew became the second group of Europeans to cross the Atlantic, where he bumped into the islands of the Carribean and, eventually, South America. It was not China, but was a satisfactory prize for the Spanish royalty. The Americas were added to the maps.

Only a few years later, Ferdinand Magellan, a Portuguese explorer, set out from Spain with a convoy of five ships. He sailed down the coast of South America, rounded its tip, and reached the Pacific Ocean. It was a bloody and dangerous trip, with many of the crew killed as they stopped to supply the ship and explore the coast. Magellan himself was killed in the Philippines. Most of his ships were lost at sea, or took on water and sank. Eventually, after 16 months, only one of Magellan's ships returned to port in Spain, its crew starving, but having made it the first ship ever to circumnavigate the globe.

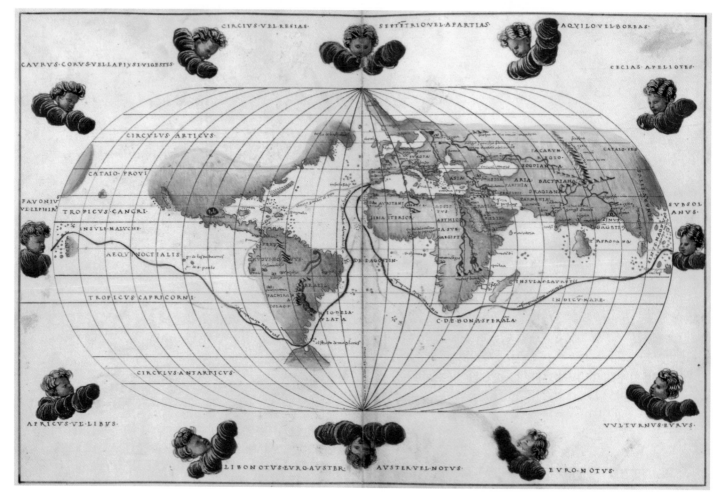

▶ A MAP OF MAGELLAN'S VOYAGE MADE IN 1544. THE FACES AROUND THE MAPS ARE THE WINDS THAT BLOW FROM EACH DIRECTION.

PHOTO: Library of Congress, Geography and Map Division

HOLDING ON TO POWER WITH MAPS AND CHARTS

Britain began building its empire in the late 1500s, and by the early 1700s it dominated the globe, with colonies and trading posts in every time zone. To keep its edge, Britain not only built the world's largest navy, but invested heavily in oceanographic exploration, the development of nautical technology, and the creation of better maps and charts. Some of the biggest challenges in nautical cartography were mapping longitude and the ocean floor.

LONGITUDE

The Greeks had come up with the idea of dividing the globe into segments using vertical (longitudinal) and horizontal (latitudinal) lines. This was to make accurate mapping easier, just as graph paper makes accurate copying easier. Latitude could be reckoned from the sun's altitude at noon, so that part was fairly easy.

Longitude was determined using what was called dead reckoning: you knew where you were yesterday, how fast you'd gone, your direction, and how long you'd traveled, and from that you could figure your new location. The problem with this method was that one error could give you the wrong location. Worse, a couple of errors could find you hopelessly lost at sea.

What you really needed to do to calculate your longitude was to determine the time using the sun. If you knew what time it was in some fixed spot, like the port you'd left from, then you could say, It's noon here, but it's 9 PM there. There are 24 hours to the day, which means we're 9/24 of the way around the world from port. With longitude settled, you could take your latitude reading, cross longitude and latitude, and find your spot on the map.

Good idea, isn't it? The only problem was knowing what time it was in the port you sailed from. Back then, finding a clock that you could carry on board, let alone check to tell London time, was impossible. Clocks relied on pendulums to keep their mechanisms turning. And in the ocean's dip and roll, pendulums flew around wildly, making it impossible to keep time.

This navigational problem was serious enough that the British Parliament put up a tremendous prize for anyone who could figure out how to measure longitude accurately. The prize was £20,000, which today would amount to millions of dollars. Inventors and scientists raced to win the prize. The winner was John Harrison, who spent 31 years refining a chronometer, a kind of clock that relied on a spring's unwinding rather than a pendulum's swing, to keep its mechanism going.

A few years later, Captain John Cook, a champion mapper and surveyor of coastlines, took his second great voyage around the world on his ship, the *Resolution*. He took with him a chronometer, which—for the first time—allowed him to find his precise longitude in the South Pacific, from the Polynesian islands to Antarctica. His coastal maps were so accurate that they were used into the 20th century.

▶ TELEGRAPH LINEMAN AT WORK ON LAND. THE WORK ON TELEGRAPH SYSTEMS AT SEA WAS EVEN MORE CHALLENGING.

PHOTO: Library of Congress, Prints & Photographs Division, LC-DIG-nclc-05117

THE OCEAN FLOOR

Mapping the surface of the ocean was one thing; mapping what lay beneath was something else altogether. And with the invention of the telegraph in the early 1800s, knowing the contours of the ocean floor took on new importance.

The telegraph was the first communication technology to link distant parts of the world; a telegraph operator in Manila could transmit a coded message, in clicks, to receivers in Montana, Paris, Odessa—anywhere else in the world the cables ran. The transoceanic cables ran for thousands of miles, but they were fragile. They could kink or break when dropped over undersea mountains. Repairing a broken cable could take days, and cost a great deal of money.

So sounding the depths—finding the bottom of the ocean to map it—became an oceanographic endeavor. The first devices were simply ropes with weights attached; the crew would lower thousands of feet of rope straight down, take the measurement, haul the weight up again, and move on. It was a tedious process, especially when the rope broke and the weight was lost. The ocean floor maps that came from these expeditions were highly inaccurate.

The first major improvement to sounding machines relied on chemistry and physics. Its inventor was William Thomson, better known as Lord Kelvin (the man who measured extremely cold temperatures). His sounding machine was ingenious: instead of rope, it used stronger piano wire. At the end of the wire there was a tube lined with a silver compound that discolored upon contact with saltwater. The extent to which seawater forced itself up the tube and discolored the silver chromate revealed the depth of the water.

Thomson's device worked well, and could be used while ships were still moving. Governments wanted all ships to carry them. Maps grew much more accurate, though there were still major debates over the existence of trenches and underwater mountain ranges. These debates would soon end, thanks to purely scientific voyages that brought a wealth of new knowledge about the seas. ∎

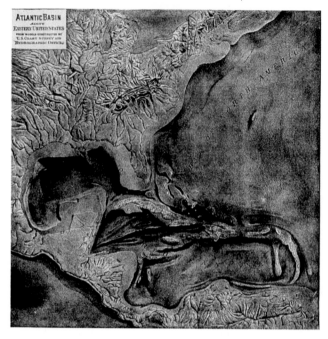

▶ **THREE-DIMENSIONAL VIEW OF THE GULF OF MEXICO. CONSTRUCTED FROM THOUSANDS OF SOUNDINGS, THIS IS THOUGHT TO BE THE FIRST 3-D SEAFLOOR MAP EVER MADE. COMPARE THIS WITH THE MODERN GEOSAT MAP ON PAGE 153.**

PHOTO: NOAA Central Library/Department of Commerce

DISCUSSION QUESTIONS

1. What are some inventions and/or improvements that made a difference in early oceanography?

2. What sorts of data, despite changing technologies over time, have people consistently collected about the ocean?

From Oceanography to MarineScience

▶ PHOTO: Nancy Marx/U.S. Fish and Wildlife Service

Our relationship to the ocean changed about 150 years ago—just yesterday, as far as history's concerned. The change came because we had new, powerful tools with which to explore the ocean. We also developed knowledge about the physics of heat, sound, and light (all kinds of waves); the ability to chemically analyze the composition of things; and the idea that living creatures are machines of sorts whose plumbing and wiring can be explored and understood. Above all, we had the understanding that the sea was made of *stuff* that was governed by physical laws and could be understood. We also had the means of getting deeper into the water, and farther into polar regions.

Oceanography today is also called marine science: it explores the chemistry, biology, geology, and physics of the waters that cover nearly three-quarters of our planet. This reading selection will give a quick overview of marine science. Like the oceans remain the least-well-explored part of our planet, there's a great deal more information about marine science, including the photos and stories of many oceanographers, to explore in libraries and online.

THE FIRST SCIENTIFIC VOYAGE

By the early 1800s, science was part of the life of the British ruling class, and they felt they must know more about the sea that held their empire together. When sailing, they had the sense of riding the back of a vast monster; they wanted to know how deep the ocean went. They wanted to know about life in its depths, if any existed (few believed it did), and to explore the creatures in distant lands that had been described as mythical curiosities. They wanted to know what the ocean was made of.

READING SELECTION
EXTENDING YOUR KNOWLEDGE

▶ DRAWING OF THE *CHALLENGER* TETHERED AT ST. PAUL'S ROCKS, AN OUTCROPPING IN THE MIDDLE OF THE ATLANTIC OCEAN.

PHOTO: NOAA/Department of Commerce

These were not practical questions. They did not link directly to empire or trade. They were asked purely in the pursuit of knowledge: a different kind of mastery of the world.

Other nations carried out dozens if not hundreds of scientific ocean voyages, and freighters trawled the oceans carrying lone scientists who'd hitched a ride. But the British scientist-explorers—well-funded, well-trained, and bold—were the best. We'll look here at the first scientific voyage, the one that set the standard for scientific oceanography for many decades.

CHALLENGER

In 1870, an energetic Scotsman named Wyville Thomson began bothering his friends at the Royal Society, Britain's club for scientists. He wanted a boat for a grand voyage of seafloor-sounding and exploration. It wasn't his first time looking for a boat. In the 1860s, Thomson had been given smaller boats and conducted wildly successful coastal sounding expeditions that revealed what it was like on the ocean floor.

He was also, by 1870, an important scientist at Edinburgh University. So in 1871 his Royal Society friends asked their Navy friends for a good ship, suitable for long voyages, and got one the following year, along with money for the voyage. The ship was a swift, steam-powered, copper-bottomed gunboat that could carry about 250 men, and its name was HMS *Challenger*.

Thomson wasted no time outfitting the ship for a grand scientific expedition, and choosing five scientists to go with him, plus a crew. He and his colleagues understood how scientifically important this voyage might be: they took out all but two guns and put in laboratories for chemistry, zoology, and botany. They stocked them well with thermometers, microscopes, sample jars, tools for chemical analysis, ledgers and notebooks, and anything else a modern lab might want. The age of photography had arrived, but they also took sketchpads, watercolors, and other art supplies for recording what they saw. The ship also carried sounding machines, hundreds of miles of twine and wire

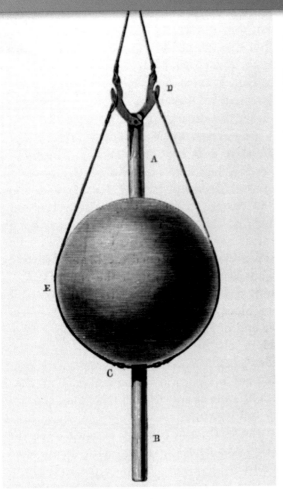

DEEP SEA SOUNDING APPARATUS THAT WAS USED ON WYVILLE'S SHIP.

PHOTO: Archival Photography by Steve Nicklas, NOS, NGS/ NOAA/Department of Commerce

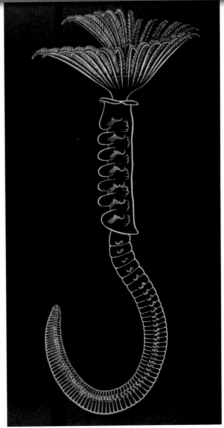

▶ **THE *CHALLENGER* EXPEDITION INSPIRED OTHER EXPLORATIONS OF THE SEA, SUCH AS THE FIRST DEEP SEA DREDGING CRUISE, CONDUCTED IN 1874, THAT YIELDED THIS DRAWING OF A TUBEWORM.**

PHOTO: Archival Photography by Steve Nicklas, NOS, NGS/NOAA/Department of Commerce

for dropping sounding weights and gathering samples of seafloor mud, and devices for measuring currents.

Over four years, *Challenger* sailed 69,000 miles, dropped anchor at 360 points, and forever put to rest the idea that the oceans were lifeless beneath the shallows. The oceans teemed with life from top to bottom, even in the cold dark depths; the *Challenger* crew caught and catalogued representatives of over 4,000 species. The scientists charted the depths as never before, discovering the 8200-meter (about 27,000-foot) deep Marianas Trench in the Pacific, and the massive Mid-Atlantic Ridge—although it must be

noted that their measurements were not precise by today's standards, and represented only 360 points scattered over all the world's oceans. They drew up from the bottoms what they called Globigerina Ooze and Pelagic Ooze, found that the ocean bottoms were covered in such oozes, and discovered what they were: sands made of the tiniest shells cast off by foraminifera.

After four years of voyaging across most of the latitudes of the world, its scientists writing and collecting and painting and analyzing the whole way, *Challenger* returned to England in triumph. It brought tens of thousands of photographs and sketches, water and sediment samples, preserved specimens, geological samples, and other artifacts. The ship was a floating museum of the world's seas. It took 20 years to publish the 50-volume account of the trip, and no serious library could do without it.

▶ SOME OF THE EARLIEST MARINE RESEARCH LABS WERE ESTABLISHED IN WOODS HOLE, MASSACHUSETTS, AND ARE STILL OPERATING TODAY.

PHOTO: NOAA/Department of Commerce

▶ CREW MEMBERS BRING A SIDESCAN SONAR SYSTEM BACK ON BOARD THE *SURVEYOR* AFTER IT WAS USED TO LOOK FOR AN OFFSHORE EXTENSION OF THE SAN ANDREAS FAULT.

PHOTO: C&GS Season's Report Jones 1965-79/NOAA/Department of Commerce

OCEANOGRAPHY TODAY

The celebrated success of *Challenger*'s expedition made others eager to go out voyaging, too. As oceanography entered the 20th century, mapping and charting remained important, and are still important today. After all, the seafloors and coasts still change. But the early 20th century also saw the birth of seaside laboratories where scientists gathered to study the oceans.

Scientists from all fields came together to study the oceans. There were biologists, who studied plants, animals, cells, fungi, the conditions in which creatures lived, and how creatures reproduced and evolved. There were chemists, who looked at what the ocean water, mud, and sea life were made of. There were physicists, who studied how sound, light, and heat traveled in the ocean; how waves formed and broke; how currents flowed; and what happened when water met air. And there were geologists, who wanted to understand the seafloor and what lay beneath it.

For the most part, these scientists designed their own exploratory equipment. Think of it: You want to see what's on the seafloor, so there's only one way to do it—go down there. Nobody else makes a vessel that can take you down, so you have to design it yourself, and

have it built, and then—try it out. This was and is extremely dangerous work. Oceanographers sometimes say that they don't like the ocean. But they do want to understand it, so they're willing to brave dampness and ice, the murk, weight, and slosh of the water, and its violence in order to study it. What follows is just a small fraction of what we've learned about the oceans in the past century.

SOUNDING WITH SOUND: PHYSICISTS MAKE BETTER MAPS

In the 1920s, a new and useful tool came to oceanography: sonar, a way to measure distances with sound. Sound waves travel well through water, and they're reflected when they hit hard objects. For the first time, scientists didn't have to drop weights and chemical tubes overboard to take soundings here and there: they could aim sound waves at the bottom of the ocean. The longer a wave took to bounce back, the farther away the bottom was.

This still wasn't simple. Think how complicated waves' reflections are when they bounce off rough, uneven surfaces, and you'll see why sonar data was tough to interpret.

MARINE CORE SAMPLING IS STILL USED TODAY. HERE, A DIVER IS TAKING A CORE SAMPLE OF A CORAL REEF TO LEARN MORE ABOUT CLIMATE CHANGE.

PHOTO: OAR/National Undersea Research Program (NURP)/NOAA/Department of Commerce

began exploring the rock beneath the mud. The urge to explore was so irresistible that—after taking a few drilled cores of rock—the geologists wanted to go all the way down to the earth's mantle to see how the rock beneath the sea had developed. Eventually they drilled about a mile into the earth under the sea.

It was a good thing, too. The rock samples they pulled up were crucial in supporting one of the most important discoveries of the 1950s: the plate tectonic theory was correct. The seafloor was not still. It was quite alive, and spreading. The chasms and rifts earlier oceanographers had found were the edges of vast plates of rock riding the mantle, which bubbled up between them and forced them apart. As the plates ground together underwater, they produced submarine earthquakes.

In these earthquakes, marine physicists identified and began to study the cause of tsunamis. Sometimes they were so intent on getting close to tsunamis that they were nearly washed away. Marine physicists devised monitoring methods to track tidal waves as they crossed the deep oceans and headed for land, and these systems have evolved into global tsunami-warning networks. They also studied how the top layers of the ocean store heat, and discovered the thermohaline circulation, a current carrying heat around the globe.

PARTICLE MEN: MARINE CHEMISTS

Chemists, too, were busy. Analysis of seawater, sediment, and shells showed how radioactive fallout from the nuclear bomb explosions of the 1940s and 1950s had made its way into ocean ecologies, and stayed. The chemists noted that carbon dioxide was dissolved in the ocean, and began to map a carbon cycle there: from air to water to shells, and back again. They worked with biologists to experiment with algae, looking for ways to use them to grow very large quantities of oils and food. This work still goes on.

It also turned out that there was a layer of ocean water that scattered sound waves before they reached the bottom. Through careful interpretation of the data, though, scientists were able to make better, more accurate maps of the bottom than ever before.

GEOLOGISTS SAY EPPUR SI MUOVE—AND YET IT MOVES

Something else amazing turned out to be true of sound waves. They could penetrate the seafloor mud. So, for the first time, marine geologists

READING SELECTION
EXTENDING YOUR KNOWLEDGE

NOT JUST JACQUES COUSTEAU: MARINE BIOLOGISTS

Biologists studied the migration habits of fish, and helped fisheries maintain their stocks; with the aid of diving techniques and scuba gear, sometimes homemade (and sometimes failure-prone), they dove into undersea expeditions. For the first time, undersea life could be explored without removal from or interference with its environment. An extraordinary diversity of marine life came to light in their work, far greater than *Challenger*'s haul of 4,000 species.

Life, it seems, can thrive in environments that seem completely inhospitable: lightless, cold, salty, and high-pressure. In the 1980s, when deep-sea vents were discovered belching boiling water and sediment into the bottom of the ocean, marine biologists discovered a world of creatures that live around these vents. These creatures use poisonous sulfur chemicals, rather than light, to feed themselves in a manner much like photosynthesis.

SAFER EXPLORATION: ROBOTS, SUBMERSIBLES, AND SATELLITES

Because the oceans are so forbidding and dangerous, there's been great interest in exploring them with robots, submersibles, and satellites. The first robots crawled the seafloor in the 1960s, and the first submersible ship carrying passengers also sank to the bottom—10.7 meters (35 feet) down—in 1964. This was *Alvin*, the "peppy little submersible." *Alvin* broke its tethering cable during a launch in 1968 and sank about 1500 meters (5000 feet). It was feared that the vessel was lost forever. When it was retrieved a year later, though, the vessel was in nearly perfect condition, and its retrieval prompted new questions for scientific study: a cheese sandwich left inside was soggy but still edible, due to the near freezing temperatures at the ocean bottom. *Alvin* still dives today, to depths of nearly 4500 meters (15,000 feet).

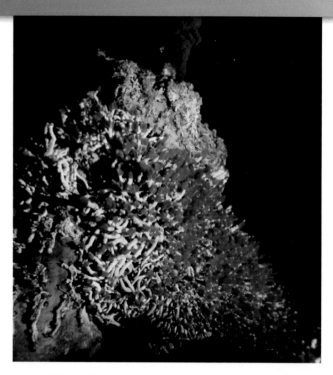

▶ **COMMUNITY OF RED TUBEWORMS LIVING ON AN UNDERSEA VENT IN THE PACIFIC OCEAN**

PHOTO: University of Washington; NOAA/OAR/OER

Today, there's a large fleet of robots and submersibles working in the ocean depths, but the idea is the same as it was in the 1960s: explore by remote control to reduce the risk to humans. The voyages of these vessels are very expensive, though—*Alvin* costs $30,000 per day to run—and they must still be taken out to sea, released, and monitored. If they're lost, they must be given up, or an expedition must be mounted to retrieve them.

Another way of looking at the oceans without risking lives is via satellite. In 1996, a new map of the ocean's surface surprised the oceanographic community with its high-resolution clarity. It had been drawn using data from GEOSAT, a satellite that bounced radio waves off the ocean to study how the earth's gravity affects the surface of the sea. GEOSAT mapped the surface of the ocean very precisely.

▶ **VIEW OF THOUSANDS OF LANTERN FISH AROUND** *ALVIN*

PHOTO: Deep East 2001, NOAA/OER

Since radio waves can't penetrate water, GEOSAT couldn't map the ocean floor. Physical oceanographers, though, knew that the closer the ocean floor is to the surface of the water, the more gravity tugs on the ocean surface— so the ocean surface dips down more over an underwater mountain range than over a marine trench. By calculating backwards from the height of the ocean surfaces, the physicists were able to put together a map of what lay beneath—one that would've taken 125 years to make by sailing a ship over the ocean's surface, and using sonar.

Much research in oceanography today has to do with understanding climate change: how currents carry heat around the globe, how oceans store heat and greenhouse gases, and how warmer oceans will affect our weather and climates. We'll turn to that science in the next lesson. ■

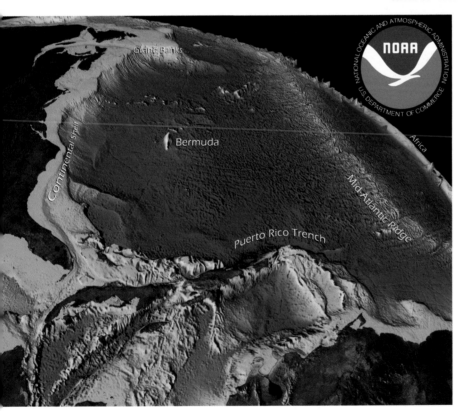

▶ **MAP OF THE NORTHWEST ATLANTIC OCEAN BASED ON DATA FROM THE GEOSAT AND ERS-1 SATELLITES**

PHOTO: NOAA

DISCUSSION QUESTIONS

1. The period described in this reading selection could be referred to as the era of scientific exploration of the seas. Explain how it differed from what came before.

2. The reading selection describes Britain's wealth in the era of its grand scientific voyages. Do you think a country must be wealthy in order to do scientific work? Why or why not?

OCEAN CURRENTS

▶ IN THE SECOND HALF OF THE 18TH CENTURY, AMERICAN STATESMAN AND SCIENTIST BENJAMIN FRANKLIN MADE SOME OF THE EARLIEST STUDIES OF OCEAN SURFACE CURRENTS.

PHOTO: NOAA/Department of Commerce

INTRODUCTION

Early explorers encountered and made use of ocean currents. These days, using temperature sensors and buoys that float at fixed depths, scientists have found that water in the ocean is continually on the move. The ocean has surface currents, produced and guided by winds. Below, currents are stirred by the earth's uneven heating. In different places, the ocean's surface waters may be warmer or cooler, saltier or fresher.

These differences in temperature and salinity power the deep ocean currents. Other factors also influence ocean currents. The earth's rotation and the configurations of coastlines and the ocean floor help shape the strength, size, and direction of ocean currents.

Such currents are essential to marine life. They carry dissolved oxygen from the atmosphere, making it available to oxygen-using creatures in the water. They also bring minerals up to the surface, where algae and other photosynthetic organisms can use them. And like currents of air in the earth's atmosphere, ocean currents help move heat around the globe.

In this lesson, you will investigate the effects of temperature, salinity, and wind on ocean currents. You will also investigate how these currents affect the climate worldwide.

OBJECTIVES FOR THIS LESSON

▸ Analyze why the water temperatures at the equator and poles differ.

▸ Investigate the effect of water temperature and salinity on density and on the movement of water.

▸ Understand that sea ice formation generates dense, salty water.

▸ Verify that ice formation in salt water increases the salinity of the remaining liquid water.

▸ Discover the difference between quantitative and qualitative observations.

▸ Investigate the effect of wind on surface currents.

▸ Locate some of the major ocean currents and analyze their effects on global climate.

▸ **MATERIALS FOR LESSON 9**

For your class
- 1 petri dish base
  Balances

For you
- 1 copy of Student Sheet 9.1: Investigating the Effect of Temperature on Ocean Currents
- 1 copy of Student Sheet 9.2: Investigating What Happens When Seawater Freezes
- 1 pair of safety goggles

For your group
- 1 copy of Inquiry Master 9.3: Mapping Ocean Surface Currents
- 1 laminated Weather and Climate World Map
- 1 plastic box with lid
  - 2 digital thermometers
  - 2 index cards
  - 1 measuring cup, 30 mL
  - 1 sheet of paper
  - 1 bottle of blue food coloring
  - 4 petri dish bases
  - 1 small cup of talcum powder
  - 5 flexible straws
  - 5 removable adhesive yellow dots
  - 2 transparency markers (blue and red)
- 1 dropper bottle of chilled food coloring solution
- 1 dropper bottle of heated food coloring solution with a test tube clamp
- 3 beakers containing room-temperature water
- 3 pieces of masking tape
- 5 containers of ice and water (labeled A through E)
- 2 empty containers (labeled F and G)
  Paper towels

GETTING STARTED

1 Look at the globe and lamp your teacher has set up. How bright do you think the light from the lamp will be at the equator? At the middle latitudes? At the poles? Discuss your predictions with your group.

2 When the clamp lamp has been turned on, look at the globe and at Figure 9.1. In a class discussion, answer the following questions:

A. Examine the equator and the poles. Where is the light brightest? Where is the globe less bright?

B. Why are some areas brighter than others?

C. Do different areas of the earth receive different amounts of light from the sun? Does each area always receive the same amount of light from the sun?

D. If areas of the earth receive less light from the sun, does it follow that they receive less heat from the sun? Why or why not?

E. Are there any other reasons why some areas of the earth might be warmer than others?

3 Listen to your teacher's explanation of differential heating of the earth. "Differential heating" means that some areas are heated more than others.

4 When your teacher invites you to, put your hands on areas of the globe that you think will be warmest and coolest. Are they? In a class discussion, answer the following questions:

A. Why are some areas of the globe warmer than others?

B. Are these the same reasons why some areas of the earth are warmer than others?

C. Why is the earth warmer at the equator than at the poles?

D. How does the sun's energy cause convection currents in the atmosphere? Do you think the convection currents at the equator are just like the ones at the Arctic Circle? Why or why not?

5 As a class, brainstorm what you know and want to know about ocean currents. Then answer this question: Are there convection currents in the ocean?

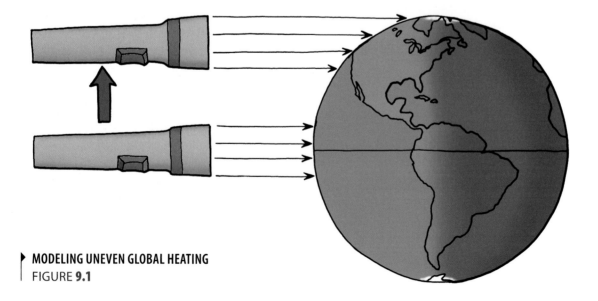

▶ MODELING UNEVEN GLOBAL HEATING
FIGURE **9.1**

INVESTIGATING THE EFFECT OF TEMPERATURE ON CURRENTS

PROCEDURE

1 Read "Ocean Currents" on page 160. How would you define a current? In this inquiry, you will investigate the effect of temperature on how water moves, and you will apply your observations to ocean currents.

2 Obtain a copy of Student Sheet 9.1: Investigating the Effect of Temperature on Ocean Currents. Read the question in the first box: How does the temperature of water affect the way water moves? Look at one set of materials. Share with your group and the class any ideas you might have for using the materials to investigate the question. Then review and discuss Procedure Steps 7–14.

3 Look at Figure 9.2 and at the plastic box your teacher has set up for this inquiry. This is a model of a deep ocean current. Your teacher will add food coloring to show how the water in the box moves. Make a prediction: As the color begins to dissolve in the water, how do you think the colored solution will move in the water?

4 If you haven't already, complete all boxes on Student Sheet 9.1 except the last one. You will describe the materials and procedures you will use, and how you will control all variables except the one you are testing, and what you will look out for and measure.

5 Review the Safety Tips with the class.

SAFETY TIPS

Be careful when you are handling the hot dropper bottle. The bottle and its contents are hot.

When your teacher hands you the hot dropper bottle of food coloring solution, the top may be unscrewed. Always handle the hot dropper bottle carefully using the test tube clamp.

Wear your safety goggles when you are using the droppers.

Do not get the food coloring solution in the bottles on your skin; it will stain.

Do not share straws; choose one group member to use the straw in this inquiry.

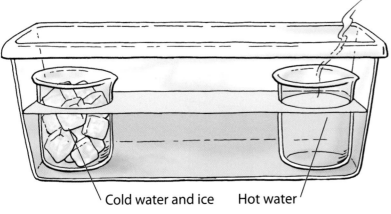

Cold water and ice   Hot water

▶ DEEP OCEAN CURRENT MODEL. DO NOT DISTURB THE WATER. VIEW THE MODEL FROM THE SIDES.
FIGURE **9.2**

Inquiry 9.1 continued

6 Pick up your materials.

7 Use the digital thermometers to measure the temperature of the water in both beakers. Record these data in the last box on Student Sheet 9.1. To make this a fair test, the temperatures should be about the same.

8 Your teacher will distribute one hot dropper bottle and one cold dropper bottle to each group. The top of the hot bottle is already unscrewed. Do not touch the glass portion of the bottle. Record the temperature of each bottle's solution.

9 Set up your investigation. As an option, you can place the cold bottle in a beaker of ice water and the hot bottle in a beaker of hot water, as shown in Figure 9.3.

10 Label your two index cards "Hot" and "Cold." Have a group member hold the index cards behind the beakers. This will make it easier to observe the results.

11 Have one group member release 10 drops of cold coloring solution into the beaker labeled as "Cold," one drop at a time. Observe the movement of the colored water from the side of the beaker. The results will be easiest to see if you are at eye level with the beaker.

Cold dropper bottle

Hot dropper bottle

▶ **SETTING UP THE TEMPERATURE LAB**
FIGURE **9.3**

12 One group member should blow gently through the bent straw across the beaker labeled "Cold."

NOTE Invert the straw so you are blowing through the long end. Rest the edge of the short end of the straw on the edge of the beaker. Do not stick the straw down into the beaker or touch it to the water.

Observe what happens from the side of the beaker. Record your observations about what happened when the cold colored water was dropped in the beaker and when air blew across the surface in the last box on your student sheet.

13 Have a group member release 10 drops of hot coloring solution into the beaker labeled as "Hot." To ensure a fair test, have the same person release the hot and cold drops. The drop dispenser should try to squeeze the droppers with the same force and speed and at the same height above the beaker in both trials. Again, observe the movement of the colored water from the side of the beaker, at eye level.

14 The same member that used the straw before should blow gently through the bent straw across the beaker labeled "Hot" (do not share straws among group members). Observe what happens from the side of the beaker. Record your observations about what happened when the hot colored water was dropped in the beaker and when air blew across the surface in the last box on your student sheet.

15 Discuss your observations with the class.

16 Clean up your workstation by setting up for the next group. Return the cold dropper bottles to the cold water bath. Your teacher will collect your group's hot dropper bottle.

READING SELECTION

BUILDING YOUR UNDERSTANDING

OCEAN CURRENTS

The waters of the ocean move in streams called currents. A current results when a fluid, either gas or a liquid, moves in a definite direction. Ocean currents form in response to many factors: heat from the sun, wind, salinity (saltiness), landmasses acting as barriers, and the rotation of the earth. Some currents are strong enough to affect the speed and direction of ships. Others bring drastic weather changes to faraway lands.

In the 18th century, Benjamin Franklin made some of the earliest studies of ocean currents. From first-hand experience, he knew that ships crossing the North Atlantic from America were either helped or hindered by a current flowing in a northeasterly direction toward Europe. Franklin investigated the temperature of the current. He found that it was warmer than the water around it. Instead of mixing with the surrounding water, the current moved along like a river within the ocean. The boundary between the warm and the cold water was sharp. Franklin discovered how sailors could use thermometers to guide their ships into the current or out of it.

Why do you think the warm water moved in a northeasterly direction? What might have caused the sharp boundary between the warm water and cold water? Think back to earlier lessons on how air moves. Try to apply some of the things you have learned about air currents to what you will learn in this lesson about ocean currents. ■

INQUIRY 9.2

EXPLORING THE EFFECT OF SALINITY ON OCEAN CURRENTS

PROCEDURE FOR PERIOD 1

1 Listen as your teacher explains the inquiry and tells you what is in the containers you will receive.

SAFETY TIP

You will be experimenting with salt water and fresh water. Tasting water is not a safe or acceptable way of telling which is which. Do not put the solutions in your mouth.

2 Your teacher will give your group three containers of ice and slushy water, labeled A, B, and E. You will also receive two empty containers, labeled F and G. As soon as you receive the containers, you will need to separate the ice from the water in containers A and B. Carefully pour the water in container A into container F while using your fingers to prevent the ice in container A from sliding out with the water. Do the same with container B, transferring the water into container G. Do not delay.

3 With your class, discuss the differences between the contents of containers A and B, when you first received them, and the contents of container E.

4 Listen to your teacher's explanation of how seawater freezes. Then answer these questions on Student Sheet 9.2: Investigating What Happens When Seawater Freezes:

A. Which freezes faster, salt water or fresh water?

B. What does salt do to the freezing point of water?

C. How does it do that?

D. What happens to salt water as it freezes? How does the ice differ from the salt water that has not frozen?

5 Dispose of the contents of container E and wipe up any water from your work area. Have a member of your group pick up a fifth container of water, labeled C, and a bottle of food coloring.

6 Listen to your teacher's explanation of what kind of saltwater mixture is in container C, and how it differs from the saltwater mixtures in containers F and G.

7 Add 10 drops of food coloring to container F. Answer the following question on Student Sheet 9.2:

E. If you were to pour very salty water into a container of mildly salty water, what would happen? Why?

HOW TRADE WINDS CAUSE UPWELLING

When you blew across the beaker in Inquiry 9.1, your breath pushed the surface water to one side. The cold colored water at the bottom rose to the top. This is how strong winds near the equator work. These winds, known as trade winds, usually blow from east to west across the tropical Pacific Ocean. Just as your breath blew water to one side in the experiment, trade winds push warmer water toward the west, where it accumulates around Indonesia. Then cold water rises from deep areas of the ocean along North and South America. The rising water brings with it rich nutrients from the ocean bottom that feed fish and marine life. This rising of cold, deep water is known as upwelling.

Sometimes the trade winds weaken, and an El Niño warming takes place. El Niño is an unusually warm flow of surface water. During an El Niño warming, there is no wind to push the warm water westward, and the cold water near the Americas cannot rise to the surface. The nutrients from the bottom of the ocean stay at the bottom. Sea organisms that depend upon these nutrients die. ■

▶ **HOW SURFACE WINDS CAUSE UPWELLING**

Inquiry 9.2 continued

8 Measure 5 mL of the colored water from container F into the 30 mL measuring cup, and slowly pour this into container C, along the side of the container. Observe container C from the side at eye level. Discuss with the class what happened, what it means, and whether it was what you expected.

NOTE Wipe the measuring cup clean with a paper towel after you have completed Step 8.

9 Listen carefully as your teacher explains the next part of the inquiry.

10 From your materials, take out three petri dish halves, masking tape, and the 30-mL measuring cup. Have a member of your group obtain a new container of salt water, labeled D.

11 Use masking tape to label the petri dishes B, D, and G. Label the sides of the dishes rather than the bottoms so the labels will be visible at all times during the experiment.

12 Measure out 15 mL of the water from container B into petri dish B using the 30-mL measuring cup. If the ice in container B has not completely melted, have a group member warm the container with his or her hands until the ice melts.

NOTE If you can see salt on the bottom of container B, stir to dissolve the salt before you measure the water into the petri dish.

Repeat this process for containers D and G and their matching petri dishes.

13 On Student Sheet 9.2, try to answer the following questions with your group:

F. What is in each of the three petri dishes?

G. Which dish do you think has the saltiest water?

H. How much more salt do you think it has than the next saltiest? Twice as much? Three times? More? Explain your reasoning.

I. What is the difference between qualitative and quantitative analysis?

14 Write your group members' names on a sheet of paper. Place your group's petri dishes on this paper in the storage area designated by your teacher. You'll leave them there so the water can evaporate. Clean up your other materials and work area.

PROCEDURE FOR PERIOD 2

1 Once the water has completely evaporated, pick up your group's petri dishes.

2 With the class, review the purpose of the investigation you are carrying out. State the questions you are trying to answer with this experiment. Discuss why you let the water evaporate from the petri dishes and what you are going to do with them next. Talk about the differences between qualitative and quantitative experiments.

3 Create a table in your science notebook in which you can record the weight of each of your petri dishes, and the weight of an empty dish. When your teacher instructs you to, go as a group with your dishes to the classroom balance. There you will find an empty petri dish. Weigh the empty petri dish and record its weight in your table. Weigh each of your petri dishes and record their weights as well. Be sure to label each with the correct letter. Let a different group member weigh each dish. 🖉

4 Return to your seat and calculate the mass of the salt in each of your dishes. To do this, subtract the weight of the empty dish from the weight of each lettered dish.

5 Record your answers to the following questions:

A. Which dish had the most salt?

B. How much saltier was the water in petri dish G than the water in petri dish D? In petri dish B? Did your answers surprise you?

C. You poured very salty water through simulated seawater. Suppose you were pouring that salty water into the ocean. Do you think it would fall all the way to the ocean floor? Why or why not?

D. Suppose the oceans warmed enough that sea ice could not form. What do you think would happen to the conveyor you read about? Is salinity the only thing that moves it?

E. Suppose the oceans cooled by 2°C. What effect do you think that would have on the conveyor? Why?

CURRENTS AND THE GREAT OCEAN CONVEYOR BELT

We think of the atmosphere and the oceans as being separate things, but you might say that motion in one drives motion in the other. Winds drive ocean currents, and ocean currents strongly affect the atmosphere and regional climates.

Wind blowing across the oceans produces surface ocean currents, which push the water along. Both these currents and the winds that drive them transfer heat from tropical areas near the equator to polar regions. Because ocean currents transfer heat from one region to another, they have a strong effect on climates.

Flowing northward along the East Coast of the United States is a tremendous warm-water current called the Gulf Stream. (Don't confuse this with the jet stream, which is a fast-moving air flow high in the atmosphere.) The Gulf Stream carries huge amounts of warm tropical water—about 15–20°C—into areas farther north. It's fairly salty water. That's because it originates from colder, salty water in the Gulf of Mexico and the very warm water tends to evaporate, leaving salt behind.

Winds steer the Gulf Stream away from the coast of North America and move it eastward toward Europe. As it moves north into chillier climates, it cools to about 11°C, releasing heat to the atmosphere off the coast of Europe. Gradually, the Gulf Stream widens and slows as it merges with the broader North Atlantic Drift. It approaches the chilly Barents Sea north of Norway, and flows along the coast of Greenland.

There in the north, as the waters cool to around 0°C, they release more heat. As the surface water freezes, it leaves salt behind, making the cold water under the ice even saltier—and denser. This dense seawater sinks to depths of as much as 3000 meters. This process is ongoing, with the Gulf Stream continually bringing new water, which grows colder and denser in the Arctic, and sinks.

The deep, dense, cold water then flows southward past Africa and all the way to the Antarctic. The earth's rotation twirls it, like cotton candy on a stick, and it flows eastward, splitting into two currents that flow north. One flows up into the Indian Ocean, and the other into the Pacific. As these currents near the equator, they warm up and rise to the surface. After looping around, they rejoin into one current, which heads back westward and eventually reaches the Atlantic Ocean, where it joins with the Gulf Stream.

This ocean current system, which results in huge energy exchanges due to differences in heat and density, is called the "ocean conveyor belt." It is estimated to take approximately 1000 years for the water in the conveyor belt to make a complete circuit. Can you trace its movement and temperature exchanges on the map?

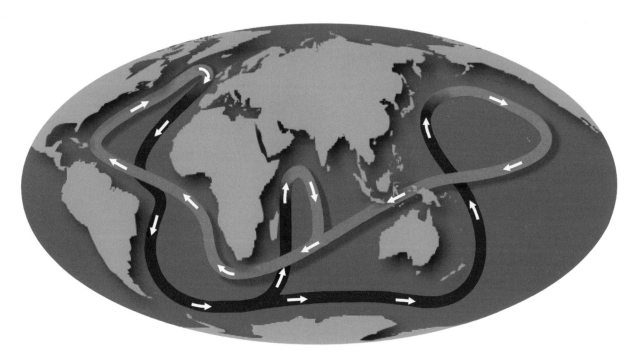

▶ **THE GLOBAL OCEAN CONVEYOR BELT MEANS THAT THE OCEAN'S WATERS ARE CONSTANTLY MOVING AROUND.**

PHOTO: Based on graphic by NOAA

What happens to the heat the Gulf Stream transfers as it travels north? Some of it goes into the atmosphere, and wind currents move the air east, over the European continent. Heat from the Gulf Stream waters warms Europe, keeping temperatures mild in winter. At the end of the last ice age, a dramatic increase in the amount of fresh water in the North Atlantic decreased the saltiness, and hence the density, of the surface waters and nearly shut down the Gulf Stream. ∎

INQUIRY 9.3

INVESTIGATING SURFACE CURRENTS

PROCEDURE

1 Read "Currents and the Great Ocean Conveyer Belt" on pages 164–165. You will investigate how wind creates surface currents. Then you will apply your observations to ocean currents on the earth.

2 Consider the following question: How do winds affect ocean waters? Discuss your ideas with the class.

3 Review Procedure Steps 4 through 8 with your teacher. With your teacher's permission, adapt the procedures to match your group's ideas as needed. Then pick up your materials.

4 Pour water into your petri dish until it is half full.

5 Use the tip of your finger to place a pinch of talcum powder on the surface of the water in your petri dish. Use only a pinch. More will interfere with your results.

SAFETY TIP

Do not share straws; each group member should use his or her own straw in this inquiry.

6 Use a straw to blow gently across the water's surface. (See Figure 9.4.) Ask other members of your group to watch the water carefully. Using a new straw each time, allow each group member to try this. What do you observe? Discuss your observations with your group.

7 Clean up by doing the following:

- Put the lid on the talcum powder. Do not throw it away.

TABLE 9.1 SOME SURFACE CURRENTS ON THE EARTH

CURRENT	HEMISPHERE	NEAREST LANDMASS OR OCEAN	DIRECTION OF FLOW	TEMPERATURE OF WATER
A. Gulf Stream	Northern	North American East Coast	North; clockwise	Warm
B. Humboldt Current (or Peru Current)	Southern	South American West Coast	North; counterclockwise	Cold
C. Antarctic Circumpolar Current	Southern	Antarctic Ocean	West to east; straight	Cold
D. California Current	Northern	U.S. West Coast	South; clockwise	Cold
E. Labrador Current	Northern	Between Greenland and Eastern Canada	South; meets Gulf Stream	Cold
F. Kuroshio Current	Northern	Japan East Coast	North; clockwise	Warm

- Pour out the water in the petri dish and wipe the dish dry.

- Do not throw away the beaker of water. The next class will use it.

- Throw away your straws. Replace them with four new straws.

- Return all equipment to the plastic box for the next class.

8 Take out your group's laminated Weather and Climate Map. Then do the following:

A. Read the descriptions in Table 9.1. Each row describes an ocean surface current caused by global winds in a certain region of the world.

B. Using the transparency markers, draw each current on your group's map. Use a red pen for warm currents and a blue pen for cold ones.

C. Label the currents with the appropriate letter (A, B, C, D, E, or F) from the table by following these steps: Write the letter on a yellow dot. Stick the dot on the map. Use a political map in your classroom, if needed, to help you locate the regions.

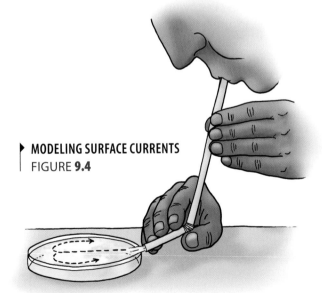

▶ **MODELING SURFACE CURRENTS**
FIGURE **9.4**

REFLECTING
ON WHAT
YOU'VE DONE

1 Answer the following questions, and then discuss your answers with the class:

A. What effect does temperature have on the movement of water?

B. What effect does density have on the movement of water?

C. How does the direction of the wind affect the direction of the surface current in your petri dish? How do you think this is similar to or different from the earth's ocean surface currents?

2 Look at the currents marked on your world map. Answer these questions:

A. What patterns do you notice in the movement of currents?

B. How do you think ocean currents affect weather globally?

C. How are the paths of the ocean surface currents you plotted on the map similar to the paths of global winds (trade winds, westerlies, easterlies)?

3 Watch as groups share the ocean currents they plotted on their world maps.

El Niño
Stirs Up the World's Weather

For fishermen living on the coast of South America in Peru, Chile, and Ecuador, business is usually good, thanks to the wind. Trade winds moving west across the Pacific Ocean, from South America's coast towards Indonesia, push warm surface waters ahead of them, piling them up. This makes the ocean surface half a meter (1.6 feet) higher and 8°C (14.4°F) warmer in Indonesia than in Ecuador, for example. On the coast of South America, only a thin layer of warm water remains, allowing cold water from a deep

▶ **FISH CAUGHT OFF THE COAST OF PERU**

PHOTO: Thom Quine/creativecommons.org

ocean current to rise to the surface. The cold current carries nutrients that nourish tiny forms of life, which are eaten by fish. These fish, in turn, are caught and sold by the fishermen.

Every three to six years or so, the trade winds weaken and stop blowing the warm Pacific water off South America westward. Surface ocean temperatures may increase from 2-3.5°C (or 3-6°F) off the coast of South America, and to depths of about 150 meters (492 feet), the water also warms. Cold water and its nutrients stay below. The fish near the surface cannot live in such warm water, and they leave to search for food elsewhere. The economies of Ecuador and other coastal countries suffer, as well as the livelihoods of fishermen and their families.

For centuries, Peruvian fishermen had noticed this decline in their fish catch every three to six years. Each time, the waters grew warmer. Christian monks named this warming "El Niño," for the boy Jesus, because the warm water came at Christmastime and, despite fewer fish, brought with it the gifts of abundant vegetation.

El Niño affects far more than the fishing industry in South America. Evaporation from the warming coastal waters causes months of rain and flooding on the west coasts of both North and South America. In other parts of the world, El Niño causes dry weather. Southern Africa, Indonesia, and parts of Australia have severe droughts during these episodes. Such big changes in the weather can be disastrous.

PEOPLE GET WATER FROM STREAMS CROSSING THE OTHERWISE DRY RIVERBED IN MOZAMBIQUE DURING A DROUGHT CAUSED BY EL NIÑO.

PHOTO: Stig Nygaard/www.flickr.com/photos/stignygaard

THE 1998 EL NIÑO CAUSED MUDSLIDES IN NORTHERN CALIFORNIA.

PHOTO: FEMA/Dave Gatley

In 1982-1983, an unusually destructive El Niño caused thousands of deaths and billions of dollars in damage around the globe. In Eastern Australia, the driest inhabited continent on earth, the worst drought in history killed livestock and devastated the economy. Many people in Indonesia died of starvation. Five hurricanes hit Tahiti in one year and left 25,000 people homeless. On the other side of the globe, southern California suffered from floods and mudslides.

READING SELECTION
EXTENDING YOUR KNOWLEDGE

▶ THIS BUOY, PART OF THE TROPICAL OCEAN AND GLOBAL ATMOSPHERE (TOGA) PROGRAM, MEASURES OCEAN TEMPERATURE AT DIFFERENT DEPTHS. THIS INFORMATION HELPS FOREWARN OF EL NIÑO EVENTS.

PHOTO: Lieutenant Mark Boland, NOAA Corps/National Oceanic and Atmospheric Administration/Department of Commerce

In response to this devastating El Niño event, scientists during the 1980s developed the Tropical Ocean and Global Atmosphere (TOGA) program to learn more about El Niño. For nearly 10 years, TOGA scientists used 70 weather buoys, moored to the ocean floor, to measure air and ocean temperatures, air pressure, seawater salinity (saltiness), water speeds, and more in the tropical Pacific Ocean. Instruments on the buoys took readings and sent them by satellite to weather centers. This information helped forecasters build computer models to predict El Niño events and how they might change weather patterns.

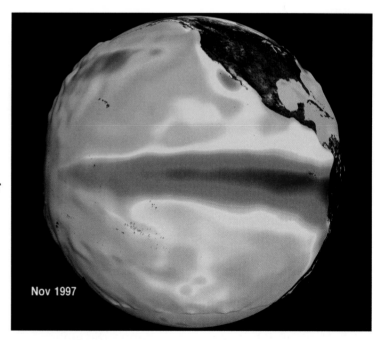

Nov 1997

▶ THIS GLOBE REPRESENTS THE APPEARANCE OF EL NIÑO'S WARM WATER IN THE EASTERN PACIFIC OCEAN IN NOVEMBER 1997

PHOTO: NASA/Goddard Space Flight Center Scientific Visualization Studio

▶ **EL NIÑO STORMS CAUSED THE RUSSIAN RIVER IN CALIFORNIA TO FLOOD IN MARCH 1998.**

PHOTO: FEMA/Dave Gatley

The El Niño of 1997–1998 turned out to be the strongest ever recorded. In the eastern Pacific, offshore ocean temperatures climbed several degrees above average. The waters were the warmest they had been in 150 years. Once again, California was inundated with floodwaters, experiencing double the normal amount of rainfall. On the other side of the Pacific, wildfires sparked and raged in the dry Indonesian forests. An airliner, lost in the thick smoke from one of the fires, crashed, and hundreds of people were killed.

However, thanks to the TOGA program, predictions helped farmers in Peru, Brazil, and Australia know what kind of weather to expect from the 1997–1998 event. When they knew that a wet season was predicted in South America, for instance, farmers there planted grains that would grow well in rainy conditions.

LA NIÑA CHANGES THINGS

El Niño is only one part of a more complex weather pattern in the tropical western Pacific Ocean that occurs every three to six years. A rise in ocean surface temperatures is followed by a general cooling of the ocean surface along the equator, an event referred to as La Niña. This is caused by strong trade winds blowing east to west.

The air surface pressure over the eastern and western Pacific Ocean also varies, swinging from high to low and back again. In normal years, warm water in the western Pacific evaporates, warming and thinning the air above it, which creates an area of low air pressure. On the eastern side of the Pacific basin, wind currents press against each other and cold air sinks, raising the air pressure on that side of the ocean. In La Niña years, surface pressure in the western Pacific near Indonesia is quite low. Think about which way the air will flow if there is high air pressure in the east and low air pressure in the west. Can you see how this might affect the temperature of the surface of the ocean?

La Niña also affects weather patterns around the world. In La Niña years, some El Niño weather patterns are reversed. The drought usually ends in Australia, and La Niña brings cold water and more fish to the coast of South America.

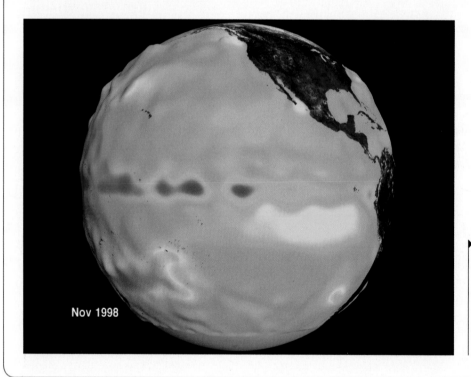

Nov 1998

▶ THIS IMAGE SHOWS THE COOLER OCEAN SURFACE TEMPERATURES OF THE 1998 LA NIÑA IN DARK BLUE ALONG THE EQUATOR.

PHOTO: NASA/Goddard Space Flight Center Scientific Visualization Studio

▶ ETHIOPIANS WAIT FOR
DONATIONS OF GRAINS, PEAS,
AND VEGETABLE OIL DURING
A SEVERE DROUGHT CAUSED
BY THE 2006–2007 EL NIÑO

PHOTO: United States Agency for
International Development (USAID)

In keeping with the cycle of every three to six years, there were also El Niño events in 2002-2003, 2006-2007, and 2009-2010. While these were not as severe as the 1997-1998 event, they did bring the characteristic changes in global conditions, causing droughts in some places, floods in others, and disturbances to marine life in the Pacific Ocean. El Niño events have been occurring for at least as long as humans have been keeping records of weather—several thousand years—and potentially for much longer. ∎

DISCUSSION QUESTIONS

1. Why is it useful to be able to predict the timing and severity of an El Niño event?

2. When scientists use models, they check different possible scenarios by varying aspects of the model, such as wind speed or temperature, and seeing how other aspects respond. If you were a TOGA scientist, what scenarios would you model? What aspects would you change and how?

BEN FRANKLIN, MATTHEW MAURY, PRINCE ALBERT, and

Rubber Duckies:
Mapping Ocean Currents

In the 18th century, Benjamin Franklin made some of the earliest studies of ocean currents. From first-hand experience, he knew that ships crossing the North Atlantic from America were either helped or hindered by a current flowing in a northeasterly direction toward Europe. Franklin investigated the temperature of the current. He found that it was warmer than the water around it. Instead of mixing with the surrounding water, the current moved along like a river within the ocean. The boundary between the warm and the cold water was sharp. Franklin discovered how sailors could use thermometers to guide their ships into the current or out of it. Not everyone wanted his advice, though. British sailors sometimes refused to take the advice of a colonist. So they sailed their own way: slowly.

In 1842, the head of the United States Naval Observatory used a different method to find currents. His name was Matthew Maury, and when he began running the Naval Observatory, he inherited the logs of former Navy ship captains. In these books the captains had recorded their locations, prevailing wind speeds, and currents. Maury studied these data and found patterns that helped him begin charting ocean currents.

Not content with the data in front of him, he convinced powerful nations around the globe that it was in all their interests to cooperate, and they sent him the data they had collected on currents. He also asked sailors on their ocean voyages to drop bottles in the ocean, each with a note giving the ship's location, the date, and a request to any finder of the bottle to return the note to Maury. He had a personal reason for making the request. When he had piloted a ship from Brazil to New York, he'd looked in vain for charts of winds and currents that would show him the fastest way there. Now he was determined to draw those charts himself.

In 1847, Maury published *Wind and Current Chart of the North Atlantic*, and it became a must-have for sailors. With Maury's book in hand, they could take advantage of winds and currents going their way and dramatically reduce the time it took to sail from one place to another. German climatologist Vladimir Köppen later improved these charts.

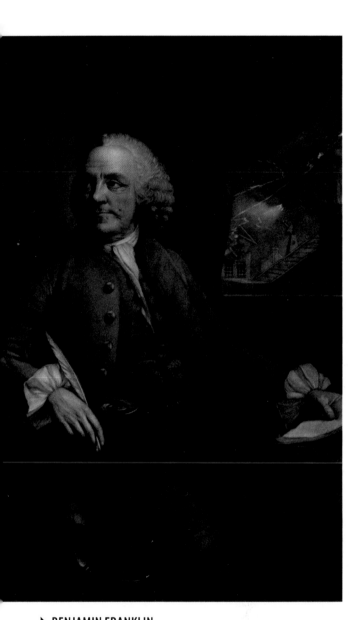

BENJAMIN FRANKLIN

PHOTO: Library of Congress, Prints & Photographs Division, LC-DIG-ppmsca-09851

MATTHEW MAURY

PHOTO: U.S. Navy file photo

▶ AN ARTIST AND A SCIENTIST IN A LABORATORY ABOARD *PRINCESS ALICE*

PHOTO: NOAA Rice Library of the National Centers for Coastal Ocean Science

The message-in-a-bottle method also helped Prince Albert of Monaco track the Gulf Stream as it approached Europe. Prince Albert became interested in oceanography as a young man after serving in the Spanish and French navies. He designed new tools for navigation and used them on his research ship, *Princess Alice*. One of his findings was that the Gulf Stream split into two parts: one that headed north, up the coast of Europe and into the Norwegian Sea, and one that headed south toward Spain and Africa.

HMS *Challenger*, which you read about in Lesson 8, also looked for currents. The method of this ship's crew was perhaps less romantic than Maury's. Wherever they stopped, they simply tied a rope to a log, tossed the log overboard, and noted the direction in which the log floated away. They also timed how fast the log pulled rope overboard, and used that measurement to gauge the speed of the current.

Today we use satellites to find and track currents—for example, to monitor the movement of a chemical spill in the water. We also release dyes into the water and track their spread. But we still toss floats overboard. Why? Because they work. Usually official research buoys and floats are tracked, but sometimes rubber duckies are used, just like those accidentally dumped into the Pacific Ocean by a cargo ship in 1992. About 29,000 of them bobbed into the ocean, and oceanographic agencies soon realized that they were ideal current trackers. By 2007, some of the rubber duckies had traveled all over the world, even across the North Pole, trapped in Arctic pack ice. ■

▶ MAURY'S CLEVER IDEA TO TRACK BOTTLES CONTAINING INFORMATION ABOUT THEIR DROP-OFF LOCATIONS ALLOWED HIM TO LEARN MORE ABOUT OCEAN CURRENTS THAN ANYONE PREVIOUSLY.

PHOTO: Mykl Roventine/creativecommons.org

DISCUSSION QUESTIONS

1. Why do you think people have been interested in mapping ocean currents since the 1800s and before?

2. If you had to use materials available in your science lab to invent a device to map currents, what would it look like?

EXPLORATION ACTIVITY: INTRODUCTION TO CLIMATE CHANGE RESEARCH

▶ A WEATHER BALLOON RELEASED OVER THE PACIFIC OCEAN CARRIES SENSORS TO MEASURE METEOROLOGICAL VARIABLES INCLUDING WIND, TEMPERATURE, AND HUMIDITY.

PHOTO: U.S. Navy photo by Mass Communication Specialist 3rd Class Kyle D. Gahlau

INTRODUCTION

This is not the first time earth's climate has changed. From the late 1500s into the 1800s, Northern European winters were harsh, and far colder than they are today. During a particularly hard 17th-century winter, the ground in parts of England froze solid a meter deep. This period of the earth's climate history is called the Little Ice Age. A deeper ice age formed Long Island, New York, 20,000 years ago. Long Island is a set of moraines, or heaps of rock and gravel pushed along the earth by advancing glaciers.

Fifty-six million years ago, the earth was warm, North America was savannah and jungle, and the glaciers were far in retreat. Two hundred million years ago, according to some researchers, Antarctica had a tropical climate, and the world's sea levels were 60 meters (200 feet) higher than they are now.

We learn how climates have changed by seeking, gathering, and analyzing data, using new techniques as they become available. Paleoclimatologists (scientists who study ancient climates) can use such data to explore, for instance, how plant species adapted to changing climates long ago. Fossil, ice core, and rock evidence of ancient climatic conditions deepens and sometimes radically changes scientific ideas of how the earth's climate has changed over time.

In this lesson you will begin a research project in which you will analyze graphs related to current climate changes. The data in these graphs was collected by research teams exploring trends in climate change. You will read about new instruments on satellites that gather climate data, giving us more vivid pictures of our atmosphere and oceans—and clues about changes to come.

OBJECTIVES FOR THIS LESSON

Discuss how scientists decide which data they want to collect.

Analyze a graph of data related to climate change for a research project.

Conduct research on the graph's data and subject.

Create and deliver an oral presentation of your findings to your class.

▶ **MATERIALS FOR LESSON 10**

For you

1 copy of Inquiry Master 10.1: Exploration Activity Scoring Rubric

1 copy of Student Sheet 10.1: Defining and Measuring the Temperature of an Area

1 copy of Student Sheet 10.2: Climate Graph (A–I)

For your group

1 digital thermometer

GETTING STARTED

1 Your group will work together to measure the temperature of an area selected by your teacher. When directed, have one student from your group pick up a thermometer.

2 Take your thermometer and Student Sheet 10.1: Defining and Measuring the Temperature of an Area to the location your teacher indicates, and measure the temperature there. It is up to your group to decide how to measure the temperature, how many measurements you will take (with a maximum of three) and where you will take them. Complete Step 1 on Student Sheet 10.1.

3 Take the measurements and record them in Step 2 on your student sheet, then return to the classroom.

4 With your group, complete Steps 3 and 4 on the student sheet. The temperature you provide for Step 3 must be a single temperature, not a range. All your group members should agree on your answer.

5 Share your results with the class and discuss the groups' results. Were some readings different from others? Why?

6 Answer these questions with your class:

A. In order to get an accurate temperature reading for the area, is it important to sample temperature in both warm and cold areas?

B. If you wanted to know whether there were convection currents in an area's air, would one temperature reading be enough?

▶ ONE MONTH'S WORTH OF SATELLITE DATA ON SEA SURFACE TEMPERATURES OVER THE ENTIRE GLOBE. RED AND YELLOW ARE WARMER TEMPERATURES, GREEN IS INTERMEDIATE, AND BLUES AND PURPLES ARE COLDER TEMPERATURES.

PHOTO: Image courtesy MODIS Ocean Group, NASA GSFC, and the University of Miami

PART 1

INTRODUCING THE EXPLORATION ACTIVITY

PROCEDURE

1 You have probably heard about climates and climate change in the news. Discuss what you've heard about climate change with the class. How do you think scientists know these things? Why might it be important to know where the data come from?

2 The Exploration Activity is an opportunity for your class to understand more about climate change data by studying data that have been interpreted as evidence of climate change. Each group will analyze a graphed set of data and present it to the class. Your group will work together to conduct research to help you understand what your graph means and why it might be important. Over the next few weeks, your group should schedule several hours for doing research, asking questions, discussing ideas, and thinking about what the graph shows and how reliably it shows that information. You will also need to think about what the graph's message means—for people, for other life forms, and for the planet.

3 Your teacher will briefly describe each of the graphs listed below.

GRAPHS OF CLIMATE CHANGE DATA

A. Carbon Dioxide Concentration at Mauna Loa Observatory, 1960–2010

B. Permafrost Temperature at Deadhorse, Alaska, 1978–2008

C. Volume of Glacier Ice, 1960–2005

D. Frost-free Days in Fairbanks, Alaska, 1904–2008

E. Changes in Carbon Dioxide Concentration in an Antarctic Ice Core over 800,000 Years

F. Sea Surface Temperatures and Hurricane Power Dissipation in the North Atlantic Ocean

G. Concentrations of Three Greenhouse Gases over 2000 Years

H. Arctic Sea Ice Extent, Annual Average, 1900–2008

I. Change in Global Mean Sea Level, 1993–2005

4 Find out from your teacher which graph your group will research in depth.

5 Obtain the appropriate version of Student Sheet 10.2 (A–I) from your teacher. It includes the graph, guidance for your research, and some questions to answer. You will frame your research by answering questions about the data used in making the graph, and a broader question about the significance of the graph. You can envision this project in parts:

1. Use the questions about the data to learn how it was collected, why it was collected in that manner, and how reliable the graph's message is.

2. Use the general research questions to understand why scientists wanted to collect such data—why the subject is important.

3. Synthesize the information collected during your research in a presentation for the class.

6 Your teacher will give you the due dates for the different components of the Exploration Activity. You will need to follow these carefully. Record the dates on Student Sheet 10.2.

Exploration Activity Part 1 continued

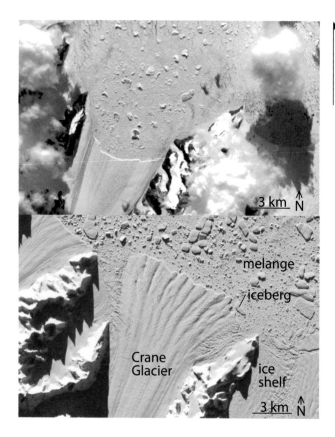

▶ THIS PAIR OF IMAGES, TAKEN BY NASA'S LANDSAT-7 SATELLITE, SHOWS THE CRANE GLACIER IN 2002 (BOTTOM) AND 2003 (TOP). THE GLACIER RETREATED AFTER THE LARSEN B ICE SHELF ON THE ANTARCTIC PENINSULA COLLAPSED.

PHOTO: NASA images by Robert Simmon based on Landsat-7 data

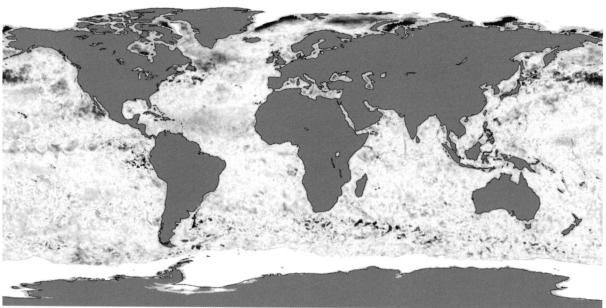

▶ A BAND OF ABNORMALLY WARM WATER, SHOWN IN ORANGE, STRETCHES ACROSS THE ATLANTIC OCEAN FROM AFRICA TO CENTRAL AMERICA. THIS REGION IS KNOWN AS HURRICANE ALLEY, THE AREA WHERE MOST HURRICANES DEVELOP OVER THE ATLANTIC.

PHOTO: Sea Surface Temperature data from the Advanced Microwave Radiometer for EOS (AMSR-E), courtesy Remote Sensing Systems/NASA Visible Earth

▸ **A GEOLOGIST STUDIES A COLLAPSED BLOCK OF PERMAFROST ON ALASKA'S ARCTIC COAST.**

PHOTO: U.S. Geological Survey/photo by Christopher Arp

▸ **EXTENT OF ARCTIC SEA ICE IN MARCH 2010**

PHOTO: NASA/Goddard Space Flight Center Scientific Visualization Studio. Blue Marble data is courtesy of Reto Stockli (NASA/GSFC)

PART 2
CONDUCTING THE RESEARCH PROCEDURE

1 As a group, carefully examine your graph and talk about what it shows. List the dependent and independent variables under Step 1 on Student Sheet 10.2.

2 Determine and record the relationship between the two variables. Look at the shape of your graph. Does it show a direct relationship (when *x* increases, *y* increases), an indirect relationship (when *x* increases, *y* decreases), or no relationship (straight vertical or horizontal line or points that are scattered widely, rather than clustered along a line)? Complete Step 2 on Student Sheet 10.2.

3 As part of your research, your group will answer the questions in Step 3 of Student Sheet 10.2. Read them over now. During your research, your group may come up with additional questions it is interested in answering. Be sure to add these to your student sheet.

4 Read Step 4 on Student Sheet 10.1, which provides a General Research Question and some topics that you could research to investigate that question. As a group, decide on one or two topics for your General Research Question. Check off the topics you select. If there is another topic you think would help your group answer its research question, record it on the student sheet. Be sure to have your teacher approve this topic before you begin your research.

Exploration Activity Part 2 continued

5 With your group, develop a plan for your research. In what order will you find your information? How will you find it? How will you record it? Outline your research plan under Step 5 on Student Sheet 10.2. As you do so, think about the number of days you have for research, and schedule your tasks so that you can complete them on time.

6 With your group, decide who will be responsible for each part of the research. For example, you might have two students answer the questions about the data, and two students work on the general research topics. Record each group member's role in Step 6 on Student Sheet 10.2. Get your teacher's approval of your research plan and group roles before you begin your research.

7 Your teacher will explain how to format a bibliography. As you work, be sure to copy down the necessary information for each source. At the end of the project, your group will need to turn in a bibliography of all the sources the group used. Your teacher may ask to see your bibliography before the final due date, to make sure you're choosing reliable sources.

8 Review Inquiry Master 10.1: Exploration Activity Scoring Rubric with the class. This is the grading rubric for the entire project.

9 Begin conducting your research. You should use a variety of printed and online sources, including magazines, newspapers, books, videos, and reliable websites. Use a minimum of five sources. If you

are not sure where to begin, ask your teacher or media specialist for help. You should also check with your teacher about the reliability of your sources before you begin using them. Record the information you find out during your research in your science notebook. 🖋

PART 3
CREATING THE PRESENTATION
PROCEDURE

1 With the class, review the rubric on Inquiry Master 10.1. Your teacher will use this rubric to assess your group's presentation.

2 Work with your partners to create a 5- to 10-minute presentation to the class. Use Inquiry Master 10.1: Exploration Activity Scoring Rubric as you create your presentation. Discuss what relationship(s) you see in the graphed data and how it relates to climate change. What will you tell the class about your general research topics? How will you turn the information you have collected into a presentation? Decide who will be responsible for creating each part, and write out what each of you will say.

3 Pictures related to the phenomena that you are studying will enhance your presentation. You might make a computer-generated presentation or a poster, or use a transparency. For instance, if you investigated data about glaciers, maps of growing or shrinking glaciers may help your audience better understand what you are talking about. Develop any visual aids you decided to incorporate into your presentation. Write captions for any images you use that explain what they are and where they came from. Be able to explain how they are related to your group's graph.

4 Your teacher will tell you when your group will deliver its presentation. Before this date, find a time outside of class to rehearse your presentation at least once. If you can find an audience to practice your presentation for, do, and ask for a critique. Use the helpful criticism to make improvements.

PART 4
PRESENTING THE RESEARCH TO THE CLASS
PROCEDURE

1 With your group, deliver your presentation to the class. Use notes if you need to, and be sure to speak clearly. Be sure all group members share in the delivery of the presentation, and that you use your visual aid. Remember, you will have no more than 10 minutes to deliver your presentation.

2 After your presentation, your group will hand in a bibliography and the visual materials you used.

3 As other groups give their presentations, make notes in your science notebook about things you find interesting about climate change. 📝

REFLECTING ON WHAT YOU'VE DONE

1 Read about the use of an array of satellites that collect environmental data in "Taking the A-Train to Make Coordinated Environmental Measurements." Discuss with the class the graph that you analyzed and whether any component of the A-train might be making measurements related to the graph's subject.

2 After all projects have been presented, write a one-page essay in your science notebook on this topic: The Most Interesting Thing I Learned From My Exploration Activity Research. Include the following in your essay:

- reasons you found this thing interesting

- how you researched it

- any challenges you ran across while researching it and how you resolved them

- any new technical or scientific ideas you learned related to the thing that interested you most (such as how a variable might have been measured, how scientists describe or name certain phenomena, and so on)

- what further research you might like to do related to the topic.

3 Write a brief paragraph explaining the thing you found most interesting in the presentations of your classmates.

READING SELECTION
EXTENDING YOUR KNOWLEDGE

Taking the  Train to Make Coordinated Environmental Measurements

In 2002, NASA and other space agencies from around the world started launching a series of satellites into orbit to monitor the earth's climates. Each of these satellites is a clever, sophisticated project, and each one makes detailed measurements of Earth as a dynamic, living planet. They monitor changes in continental glaciers and sea ice, trends in atmospheric temperature change, fluctuations in forest cover and soil moisture, concentrations of atmospheric particles, and changes in cloud altitudes, ozone thickness, sea surface temperature, ocean circulation, and much more. Those sets of data provide information on Earth's climate trends and human impact on climate.

In designing them, NASA's scientists intended for each satellite to travel by itself. But two environment-observing satellites, Aqua and Aura, were already orbiting near each other when engineers decided to use Aura's orbit for a third satellite instead of designing a new orbit. It turned out that engineers on two other projects

▶ **AN ARTIST'S CONCEPTION OF THE A-TRAIN SATELLITES IN 2014**
PHOTO: NASA

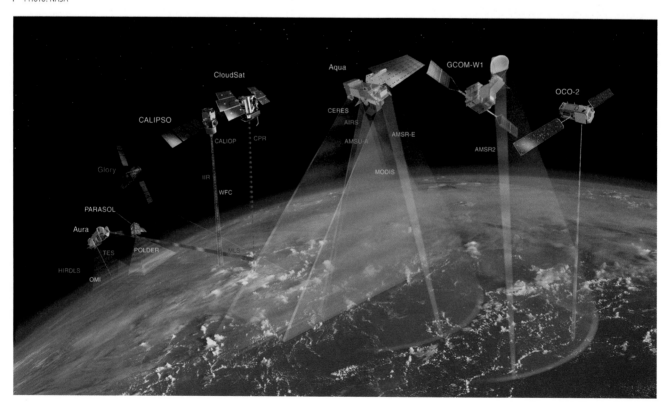

▶ IMAGE OF HURRICANE BILL (2009) USING DATA COLLECTED BY *AQUA*, AS WELL AS OTHER SATELLITES

PHOTO: NASA

had asked to use Aura's orbit, too. Suddenly NASA had a whole formation of environmental satellites moving together.

Following one another over the equator, this multiple-satellite program was dubbed the A-train, after a jazz tune about a subway line in New York City ("Take the A-Train"). In this case, the "A" refers to "afternoon." The satellites cross the equator at about 1:30 pm each day. Strictly speaking, the A-train satellites don't function exactly like a train, since they don't follow each other in a straight line and do not depend upon an engine. They do, however, move in a coordinated formation and collect data in sequence. While each satellite

was designed to study aspects of different earth systems, the engineers soon found that there was a great advantage to having them travel together.

Because these satellites pass over the same spots on the earth within minutes of each other, they reveal many things about a specific place's environment at a specific time. This turned out to be useful when the area of ice cover in the Arctic melted to record lows in 2007. A review of the satellites' data showed several kinds of strange atmospheric events over the Arctic as the ice melted—unusually strong winds and sunny skies, for instance—and gave new insight into why polar ice melts.

READING SELECTION
EXTENDING YOUR KNOWLEDGE

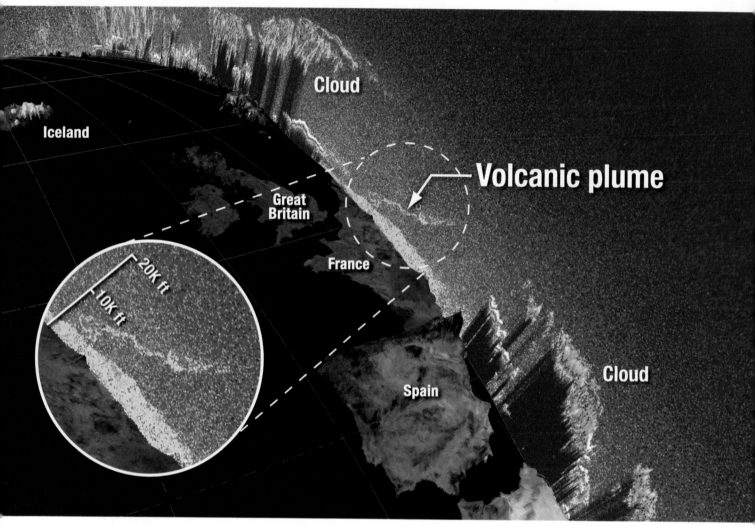

▶ IMAGE SHOWING DATA COLLECTED BY *CALIPSO* ON THE VOLCANO PLUME FROM THE 2010 ERUPTION OF ICELAND'S EYJAFJALLAJÖKULL VOLCANO.

PHOTO: NASA/Kurt Severance and Tim Marvel

MEMBERS OF THE A-TRAIN

The satellite *Aqua*, launched in 2002, crosses the equator first. "Aqua" means "water" in Latin, and the name fits. *Aqua* carries six instruments from Brazil, Japan, and the U.S. that monitor all parts of the water cycle: evaporation, precipitation, variation in land and sea ice, ocean surface temperatures, soil moisture, and snow cover. *Aqua* also monitors the amount of plant life in water and on land; temperatures of land, air, and water; and how the radiation energy from the sun fluctuates. *Aqua* tells about the earth's energy balance (the amount of solar radiation reaching the earth versus the amount of energy escaping from the earth into space), what role clouds play in blocking and capturing energy, and whether the water cycle is speeding up.

Why is the speed of the water cycle important? One reason is that a faster, heavier-duty water cycle means more flooding. Warmer, higher seas and warm air mean lots of evaporation and heavy clouds. When those clouds reach coastal areas, they bring more rain than those areas normally see. When many rainstorms follow one right after another the land does not have time to dry out, and this can have serious consequences for organisms living in the area.

Several of the satellites that followed *Aqua* into space contain instruments for studying clouds and tiny particles in the earth's atmosphere called aerosols. Aerosols are extremely important because they can both trap and reflect solar radiation. They affect how much energy reaches the earth's surface, and how much energy stays in the atmosphere.

CloudSat, a joint project of the United States and Canada, was launched in 2006. *CloudSat* studies clouds and the role they play in the earth's climate. We know that clouds reflect solar radiation and trap radiation that is rising from the earth, but what's their overall effect? What effect do clouds have on the earth's temperature? And how is the earth's warming and cooling affected when clouds overlap in layers? *CloudSat* uses microwave radar to scan for clouds. Microwaves are extremely small. This radar can detect 90 percent of all water clouds and 80 percent of ice clouds in the atmosphere. By coordinating *CloudSat* data with other environmental data, scientists hope to understand more about how clouds affect the earth's climate and weather.

Following close behind *CloudSat* is *CALIPSO*, also launched in 2006. *CALIPSO* is a joint effort of France and the United States. This satellite carries three instruments that collect data on how clouds and aerosols interact. We already know that aerosols can affect air quality. For instance, aerosols like soot can make breathing difficult for people with lung diseases. But how do aerosols affect clouds? Do they turn clouds into tighter "lids" on the planet, keeping in the heat? *CALIPSO*'s constant watch on aerosol–cloud interactions, coordinated with data from other environment-monitoring satellites, should help answer these questions.

PARASOL is another French A-train satellite, launched in 2004. This satellite samples and measures light reflected from the earth's surface and atmosphere, using the light's wavelengths to map and identify aerosols and clouds. One important thing *PARASOL* does is monitor polarized light, or light waves that have been forced to travel in a single plane. Small molecules can polarize the light that bounces off of them. The way that a molecule polarizes light can tell us something about how the molecule is put together, and helps us identify it. This "molecular signature" can help determine whether the molecule is man-made or natural. *PARASOL*, in other words, helps scientists determine what proportion of the aerosols in the atmosphere are man-made. That's important for policymakers. If our aerosols contribute to climate change, we need to figure out how to avoid releasing them into the atmosphere.

PARASOL originally flew in tight formation with the A-train, but dropped to a lower orbit in 2009. It will continue to fly under the neighborhood of the A-train now and then, but its orbit is decaying, meaning it is spiraling down towards earth. NASA had hoped to replace *PARASOL* with *Glory*, another satellite meant to study polarized light in the atmosphere. Unfortunately, *Glory* failed to launch successfully in 2011, and there is no planned replacement.

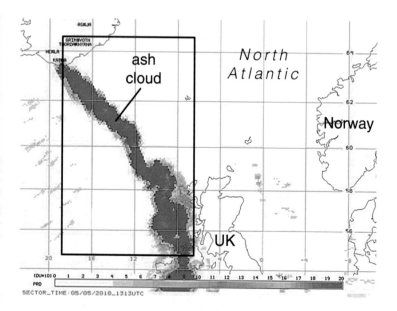

▶ DATA COLLECTED BY *AURA* SHOWS A PLUME OF SULFUR DIOXIDE FROM EYJAFJALLAJÖKULL VOLCANO.

PHOTO: NASA

Aura, also launched in 2004, flies 15 minutes behind *Aqua*. It's something of a pollution-studying superstar, carrying instruments that monitor both pollution levels in the atmosphere and how man-made pollutants travel through the atmosphere. It is important for scientists to be able to distinguish man-made pollutants from natural ones in the atmosphere in order to understand how human activities are affecting air quality.

Aura also monitors the ozone hole to check the progress of ozone-hole recovery. Ozone is a three-oxygen molecule; a layer of ozone is found in the stratosphere. It acts as a shield, and absorbs most of the damaging ultraviolet radiation that hits the earth. Humans have made a huge hole in this shield with chemicals that are popular refrigerants and spray propellants. These chemicals, called chlorofluorocarbons (CFCs), are harmless to humans, and we've used millions of tons of them, unaware of the damage they were

causing. The use of CFCs is now largely banned, but their effects can still be measured.

The A-train's newest members are GCOM-W1 and OCO-2. The Global Change Observation Mission—Water (GCOM-W1) is a Japanese satellite that will collect important data about how water moves around the globe. Why are scientists interested in this? Think about strange weather events of the past few years: unusually heavy snowfalls, droughts in areas that are usually wet, an unusual number of typhoons and strong hurricanes. GCOM-W1 can detect radio waves from the earth's surface and atmosphere, monitoring where water is and its temperature. All objects naturally emit radio waves. The waves' strength tells something about how much moisture the object has and its temperature. By scanning the radio waves emitted by a section of land, the satellite can tell how much water the land holds and how warm that water is. That data can be matched to data about storms, droughts, and other water cycle events in that location, and scientists can see what patterns emerge.

The last part of the satellite formation, the Orbiting Carbon Observatory-2 (OCO-2) will measure the atmosphere's concentration of carbon dioxide, a greenhouse gas emitted in large quantities when fossil fuels like oil and coal are burned. How much carbon dioxide humans put into the atmosphere can be estimated by adding up the amount of fossil fuel we use. But we don't know which countries, or which areas, are adding the most carbon dioxide to the atmosphere, nor do we know how effective forests and grasslands are at removing it. OCO-2's instruments will map where carbon dioxide is being pumped into the atmosphere and where it's being taken out. By scanning columns of air from satellite height all the way down to the planet's surface, OCO-2 will see where carbon dioxide is concentrated throughout the earth's atmosphere.

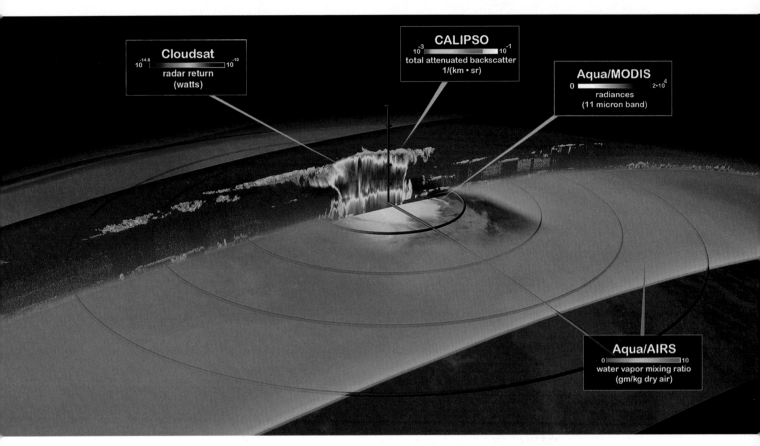

Cloudsat
$10^{-14.8}$ 10^{-10}
radar return
(watts)

CALIPSO
10^{-3} 10^{-1}
total attenuated backscatter
$1/(km \cdot sr)$

Aqua/MODIS
0 $2 \cdot 10^{4}$
radiances
(11 micron band)

Aqua/AIRS
0 10
water vapor mixing ratio
(gm/kg dry air)

▶ **TROPICAL STORM DEBBY (2006) WAS OBSERVED BY FOUR DIFFERENT A-TRAIN INSTRUMENTS**

PHOTO: NASA

The A-train can collect data on carbon dioxide concentrations, map clouds and aerosols, check on where water is and how fast it's cycling, monitor and map greenhouse gases and the ozone hole, and monitor energy entering and leaving the earth. It's like having a team of doctors and scientists all standing around one patient, Earth, monitoring different parts, noticing changes and trends, working together to figure out cause and effect in the patient's health. The hope is that the A-train's data will give a new, richer picture of our home, help us understand its climates, and help us keep it healthy. ■

 DISCUSSION QUESTIONS

1. What are the advantages of so many countries working together on satellite programs? Are there any reasons why they might not want to do so?

2. How do the A-train satellites help prevent and deal with global environmental problems?

EARTH'S CLIMATE ZONES

▶ SPECTACLED EIDERS, A THREATENED BIRD SPECIES, SPEND THE WINTER CLUSTERED IN AN OPENING IN THE ICE PACK ON THE BERING SEA. THEY MAY HAVE TO ADAPT TO CHANGES IN THE ICE PACK FROM GLOBAL WARMING.

PHOTO: William Larned/U.S. Fish and Wildlife Service

INTRODUCTION

The history of climatology is short, with major scientific work beginning only about a hundred or so years ago. In that time, scientists have learned to map climates. They have also learned that climates change and change greatly. Scientists have found that the physical evidence left behind by climate change can be used to build pictures of what an area's past climates were like. These pictures help scientists imagine what happened to life and landforms in the area as the climate changed.

In the last few decades, governments and scientists all around the world have collaborated to learn a great deal more about how climates are made and why they change. This work is being done because of mounting evidence that earth's climates are changing rapidly, and that the consequences for life on earth may be serious.

This lesson begins with the first definitions of climate zones, made in the 19th century, and introduces you to the current work of climate scientists. You'll take a long look ahead at how your region's climate may change in the next hundred years, and longer looks into the past at climates that existed 55 million years ago. You will participate in this work, using what you know about climate and weather to make recommendations for how people in your area can prepare for climate change. You will also learn to interpret evidence of climate change that happened six epochs ago.

OBJECTIVES FOR THIS LESSON

Examine factors that influence climate.

Examine natural climate zones on earth.

Study federal climate change projections for a region of the United States.

Develop climate change–related policy recommendations for the governor of your state.

Investigate the use of plant fossil data as an indicator of past climate.

▶ MATERIALS FOR LESSON 11

For you

1	copy of Student Sheet 11.1: World Climate Zones
1	copy of Student Sheet 11.2: Smooth or Jagged?
1	colored pencil

For your group

| 1 | set of Leaf Fossil cards (A and B) |

GETTING STARTED

1 With your class, review what you know about weather and climate.

2 Read "Climate Classification System" on pages 195–197. What climate zone do you live in?

3 Examine Figure 11.1, which shows the world's climate zones, below. Find the region in which you live.

4 On Student Sheet 11.1: World Climate Zones, outline your region and shade it in.

5 Discuss these questions with your class:

A. What conditions define your climate zone?

B. What nearby areas are also in your climate zone?

C. How are two neighboring climate zones different from yours?

6 Use Figure 11.1 to find all the other regions that share your area's climate type. Outline each of them on Student Sheet 11.1 and shade them in.

7 Record your answers to the following questions on the student sheet:

A. How many regions share your area's climate type?

B. What countries have regions of your area's climate type? How are those places like your local area?

C. What percentage of the world's landmass, approximately, shares your area's climate type? Try to estimate to the nearest 10%.

8 Use the map in Figure 11.1 and what you've learned about climate zones to discuss with the class what appears to influence a region's climate type most. Is it latitude? Proximity (closeness) to oceans? The presence of mountains? Other factors? A combination of factors?

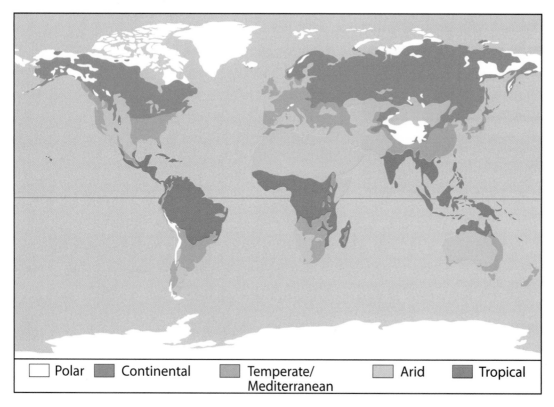

Polar ☐ Continental ▢ Temperate/ Mediterranean ▢ Arid ▢ Tropical ▢

▶ THIS MAP SHOWS THE KÖPPEN SYSTEM FOR CLASSIFYING CLIMATE INTO FIVE MAIN ZONES.
FIGURE **11.1**

PHOTO: Based on graphic by NOAA

CLIMATE CLASSIFICATION SYSTEM

There are many ways of classifying the earth's climates, but the most widely used methods are based on the work of German climatologist Vladimir Köppen. Between 1884 and 1936, Köppen examined plant and animal life in various climatic conditions (e.g. wet and dry, warm and cold) around the globe. He believed that looking at life—especially how it responds to and tolerates various conditions—was the best way of understanding regional climates.

Köppen divided the planet into five main zones: tropical, arid, temperate, continental, and polar. Within each of these zones, he found that there were variations; for example, some tropical areas have more steady temperatures than others. Aware of these differences, Köppen divided each zone further into sub-zones.

The tropical zone has a high average temperature year-round, 18°C or higher, and is wet. Although you might think of the tropics as being equatorial, not all tropical zones are near the equator. Nor are they all jungle. Tropical areas also include places such as Hilo, Hawaii; Singapore; Miami, Florida; Cairns, Australia; Rio de Janeiro, Brazil; and Jakarta, Indonesia. The tropical zone has three sub-zones: areas that have a dry season, areas with monsoon (torrential rain) seasons, and hot, damp areas that have more or less consistent rainfall all year round.

► **LUSH, WET RAINFOREST IS CHARACTERISTIC OF THE TROPICAL ZONE.**

PHOTO: Randolph Femmer/life.nbii.gov

READING SELECTION

BUILDING YOUR UNDERSTANDING

continued

▶ **DESERT PLANTS ARE HIGHLY SPECIALIZED TO DEAL WITH DRY CONDITIONS IN THE ARID ZONE.**

PHOTO: George Gentry/U.S. Fish and Wildlife Service

The arid, or dry, zone includes places such as North Africa; Dubai, United Arab Emirates; Tijuana, Mexico; and Yuma, Arizona. Areas in this zone are extremely dry, receiving little rainfall. This creates a challenge for plants, which regularly lose water through evaporation and transpiration. Some plants with specialized features for water conservation, such as cacti, grow in these regions. There are two types of arid zones: steppes, which are grasslands that may have a few damp months each year, and deserts. It's interesting to note that not all dry areas are hot. Arid zones have a wide range of temperatures. The Gobi Desert in China, for instance, sometimes has frost.

The temperate zone is just as its name implies: not very cool or very hot. This zone has an average temperature range between -3–10°C and includes Mediterranean countries, the Pacific Northwest, and much of the eastern United States. Temperate zone vegetation is diverse but includes deciduous trees which shed their leaves annually to adapt to winter conditions. The three main types of temperate zones are those with dry summers, those with dry winters, and those that are damp and rainy year round.

A larger area, the continental zone, has warm or hot summers and very cold winters. It has deciduous vegetation that can survive these large annual fluctuations in conditions. Continental areas are often far from oceans, located in the interior of continents. Chicago, Illinois; Belgrade, Serbia; and Toronto, Canada are all located in continental zones. In Chicago, for instance, it isn't unusual to find summer temperatures near 38°C and winter temperatures below -18°C. Some continental areas tend to be dry in the winter, and others in summer.

▶ **ANIMALS THAT LIVE IN THE POLAR ZONE, LIKE THIS POLAR BEAR, HAVE ADAPTATIONS SUCH AS THICK FUR TO SURVIVE IN THE COLD TEMPERATURES.**

PHOTO: George Gentry/U.S. Fish and Wildlife Service

The fifth climate zone, the polar zone, has average monthly temperatures below 10°C all year. It includes the tundra and ice cap regions of the world, with the small towns of Barrow, Alaska; Provideniya, Russia; and Nuuk, Greenland represented in this zone. Neither tundra nor ice cap can support trees, but the tundra does have plant life: shrubs, mosses, lichens, and grasses. Still, the land is dry, and not much can live there.

Ice cap regions are even colder, with average year-round temperatures below -18°C. These areas are as dry as deserts. Astonishingly, life can be found in the form of cold-hardy bacteria, plankton, algae, fungi, and even small invertebrates.

Which climate zone would you prefer to live in? ■

INQUIRY

INQUIRY

PROJECTING CLIMATE CHANGE

PROCEDURE

1 Listen carefully as your teacher introduces the inquiry and explains climate modeling.

2 Examine the climate change projections your teacher displays, and discuss what you notice with the class.

3 Answer the following questions as a class:

A. How could you find out whether the scientists who made these graphs are credible scientists whose work you can trust?

B. Why is there a range of potential changes shown in these graphs rather than a single projection?

4 Suppose that your group is advising the governor on climate change. Given your knowledge of weather, climate, and the climate projections you have just discussed, what measures would you suggest the governor take to help your state prepare for projected climate changes? Be sure to make recommendations for several areas of your state: waterfront areas, cities, farmland, mountainous areas, and so on. If there are recommendations that can apply to the whole state, make them, too. Record your recommendations in your science notebook. ✏️

5 When all groups have written their recommendations, share yours with the class.

CLIMATE SIGNALS: SMOOTH OR JAGGED LEAVES

PROCEDURE

1 In this inquiry, which was developed by the Smithsonian Center for Education and Museum Studies, you will examine photographs of actual leaf fossils collected at two dig sites in the Bighorn Basin in Wyoming and calculate the temperature change that occurred there. Your work will be very similar to that done at the Bighorn Basin, except that you won't dig for the fossils. Read "Prehistoric Climate Change" on pages 200-203.

2 Pick up and organize the sets of cards (A and B) for this activity. Half of your group will start with the cards from Set A, and the other half from Set B. Arrange the cards from your set in numerical order and examine each fossil card carefully to categorize each leaf as having a smooth or jagged (toothed) edge. Look for regular patterns around the edges, and don't include tears or insect bites.

3 Record the data on Student Sheet 11.2: Smooth or Jagged?

4 Exchange cards with the other half of your group and complete your analysis. Complete Steps 3-5 on Student Sheet 11.2.

5 Complete Step 6 of the student sheet, and be prepared to share your data and your theories with the class.

6 Read "Causes of Climate Change: Natural and Manmade," on pages 204–209.

REFLECTING
ON WHAT
YOU'VE DONE

1 Answer these questions in your science notebook, then discuss them with the class:

A. What kinds of climatic conditions distinguish the world's climate zones from one another?

B. What is the difference between "climate" and "weather"?

C. Why don't all future climate projections agree with each other?

D. How could you find out about how the climate in your area has changed in the last 75 years? 125 years? 10,000 years?

PREHISTORIC
CLIMATE CHANGE

About a century ago, botanists started noticing that most tree species living in warm climates have leaves with smooth edges, while most tree species in cooler places have toothed, or jagged, edges.

It's no accident that botanists noticed this connection between leaf shapes and temperature. Natural historians—early scientists who investigated nature by observing it closely with their own senses—went exploring in the 17th, 18th, and 19th centuries with sketchbook and pencil in hand, drawing carefully what they'd seen, and recording their observations about the surrounding environment. They collected samples to investigate closely at home. They produced beautiful drawings and watercolors of variations in plant life and leaf shape. Modern botanists who study the shapes of leaves and their connection to climate are following in the footsteps of these early naturalists.

So why are leaf shapes so sensitive to temperature? The thin tissue at the edges of jagged leaves matures quickly, allowing leaves to start photosynthesizing. Scientists think that starting photosynthesis earlier in the spring may be an important advantage in places with a cool, short growing season. Smooth-edged leaves appear to lose less water to the atmosphere, and thus are advantageous in warmer climates where water loss is often a problem for plants.

A relationship between temperature and the proportion of tree species with smooth-edged leaves has been observed in forests all over the world. It can be expressed mathematically in an equation that paleobotanists apply to fossil leaves to calculate the temperature when they were alive. (*Paleo-* means ancient; paleobotanists study fossil plants and the plant life of ancient earth.) The equation is used to estimate temperatures in the distant past by looking at the proportion of fossil leaf species that have smooth edges. And a series of leaf fossils from the same location, but from different times in the earth's history, can give paleobotanists an understanding of how the climate of that spot changed over time.

The trick is finding leaf fossils. Normally, leaves decay instead of fossilizing. Under the right conditions, though, they leave vivid fossil imprints in the hardening mud in which they've been entombed.

▶ **19TH CENTURY WATERCOLOR DRAWING**

PHOTO: Library of Congress, Prints & Photographs Division, LC-DIG-ppmsca-22947

TOOTHED LEAVES, LIKE THOSE SEEN HERE OF A YELLOW BIRCH, ARE MORE COMMON IN COOLER CLIMATES.

PHOTO: Robert H. Mohlenbrock @ USDA-NRCS PLANTS Database / USDA NRCS. 1995. Northeast wetland flora: Field office guide to plant species. Northeast National Technical Center, Chester.

WYOMING'S BIGHORN BASIN

We talk a lot about global warming these days, but this is not the first time the earth's climates have changed. Earth's geologic history is marked by swings in temperature large enough to make North America glacial and subtropical by turns, the sea level rising and falling as ice sheets melted and built up again.

During the early Cenozoic Era, 55 to 65 million years ago, Wyoming was not at all the rocky, dry, blizzard-prone ranchland we know today. Instead, it was a swampy lowland inhabited by alligators, soft-shelled turtles, and the ancient relatives of palm trees and bald cypresses. The environment was warm; frost was a rare event, and the entire planet was ice-free. Alligators lived north of the Arctic circle, and palm trees grew as far north as Canada.

The mountains in Wyoming—the Rockies and other ranges—were still growing. As the mountains rose, sediment eroded from their sides, burying dead plants and animals. The sediment eventually turned to layers of sedimentary rock. The Bighorn Basin is a wide valley between Rocky Mountain ranges that accumulated up to 3800 meters (more than 12,000 feet) of sedimentary rock during this time. And for more than a century, paleontologists—the fossil hunters and scholars of ancient earth—have been working in that barren landscape, digging up the past.

Scientists have found hundreds of types of fossil plants and mammals in rocks from the Bighorn Basin. Many of these are about 56 million years old (10 million years after the extinction of the dinosaurs). The fossil mammals include the earliest horses and primates. These early horses were very small—about the size of a house cat. The early primates, members of the same group of mammals to which humans belong, were lemur-like tree climbers that ate fruit and leaves.

LEAF FOSSIL FROM A RIDGE IN SOUTHERN WYOMING

PHOTO: National Park Service

READING SELECTION
EXTENDING YOUR KNOWLEDGE

▶ **FIELD CREW LOOKING FOR FOSSILS IN THE BIGHORN BASIN**

PHOTO: Scott Wing, National Museum of Natural History, Smithsonian Institution

SCOTT WING'S DISCOVERY

Paleobotanist Scott Wing has been working in the Bighorn Basin for more than 20 years. He's a well-known scientist who's often helped others understand the relationships between leaf shapes and climate. He's well aware of how rich the Bighorn Basin is in fossils; hundreds of thousands of fossils have been dug from its rocks. But until 2003, one thing was missing from the fossil record in the Basin and indeed from everywhere else: plant life from an important period in the earth's history. The period was called the Paleocene-Eocene Thermal Maximum (PETM), named for the first two time periods of the early Cenozoic.

When he found no fossil leaves from this period in the vast areas where geologic clues told him he ought to find them, Wing went over the terrain again, and again, and again, failing every time to find what he was after. But in 2003, the rocks proved him right. Wing found a single leaf fossil at the edge of what would be a trove of over two thousand fossil leaves, and the discovery, after all that time hunting in that arid basin, made him sit down and cry.

Wing dated the rocks containing the leaf fossils by studying the layers of rock, and with reference to nearby animal fossils of known age. Comparing fossil leaves from before the PETM with those he found in the PETM, Wing discovered that species with smooth leaves became much more common during the event. The fossils provided evidence that, 56 million years ago, the climate warmed suddenly by about 5° Celsius.

Wing's results are consistent with temperatures reconstructed from fossils of marine microorganisms, called foraminifera. The chemistry of foram shells of the same age as Wing's leaf fossils also show about 5°C of warming 56 million years ago, indicating that the warming was a global event. Both the foram data and the Wyoming leaf data indicate that the global warming took place in less than 20,000 years (a blink of an eye in geologic time) and lasted another 100,000 years before cooling set in.

While five degrees might sound trivial, that much warming, that fast, was cataclysmic. Species disappeared. Climates changed so much that places became unrecognizable in terms of plant and animal species. In the Bighorn Basin, where swamp groves similar to present-day coastal North Carolina once had been, there

▶ **FOSSIL PALM FROND**

PHOTO: Scott Wing, National Museum of Natural History, Smithsonian Institution later grew plants that belonged to a dry, tropical climate: mimosa and palms, trees like acacias, and others whose relatives now live in Mexico.

What caused this dramatic warming event? One clue from studies of ocean sediments from 56 million years ago is that a large amount of carbon dioxide entered into the ocean and atmosphere. Scientists have several hypotheses about where the carbon dioxide came from. One idea is that increased volcanic activity released carbon dioxide from organic-rich sediments. Another idea is that methane deposits in the ocean floor may have been released to the atmosphere, causing warming (methane is a strong greenhouse gas), and rapidly converted to carbon dioxide, which would also have caused warming.

Scientists are interested in this long-ago period of global warming because it helps us understand how the earth's climate responds to greenhouse gases like carbon dioxide, and can help us predict the effect of our own carbon dioxide and methane emissions on future climate. If we burn the rest of our fossil-fuel reserves in the next few hundred years, Dr. Wing says, we will release carbon into the atmosphere more than ten times faster than it was released during the 10,000 years of rapid warming at the end of the Paleocene epoch. That would mean the earth would warm even faster now than it did back then. Will our plants survive, and will warm-climate plants move northward? Will the world's heterotrophs—including humans—have enough to eat?

Finding and interpreting these clues to the earth's past teaches us about what happened before, and helps us guess what might reasonably happen again. But the future will test our hypotheses. Wing cautions, "The central truth about predicting global change is that the systems we are predicting are very complicated and very interconnected. We can make all the supercomputer models we want, but we won't really know how well these models predict the future until we get there . . . The great advantage of the fossil record is that we can study events that have already happened and work out the links . . . It seems that the least we can do, now that we are modifying the planet we live on, is to read the operator's manual written in its rocks." ∎

DISCUSSION QUESTIONS

1. Explain why finding a single leaf fossil brought a paleobotanist to tears.

2. Why is it particularly important to see how plants, rather than animals, adapt to climate change?

CAUSES of CLIMATE CHANGE: Natural and Manmade

Imagine: you're in your backyard, gardening. The tomatoes are not ripening at all: a total loss. It seems to you that there's been a lot more rain than there was last year, and the summer was cooler than usual. How do you know if your perceptions are accurate? And—more important for your garden planning—how do you know whether this is a fluke or part of a trend towards cooler, rainier weather? Will your tomato plants survive next year, or should you plant vegetables that don't need hot, sunny summers?

▶ **YOU WERE HOPING FOR SOMETHING SWEETER THAN A GREEN TOMATO!**

PHOTO: Angela Rucker/U. S. Agency for International Development

You might go online to check weather records, look for trends, and read climate scientists' reports on your area. It's possible to do that because for over a century, scientists have kept careful track of the weather all over the world, monitoring temperature, precipitation, winds, and humidity; noting when the ground freezes, the first and last frosts, and many other weather variables. And thousands of meteorologists and climate scientists are working to understand what these data mean.

But imagine that you live in 1816, before the routine collection of accurate weather measurements. There's no such thing as a climatologist. You know very little about what air is made of, or where the sky stops, or what might lie beyond it. No one knows much about what makes rain fall, and thermometers are expensive and hard to use. Still, you know that summer is supposed to be warm. Instead, it's *cold*: alarmingly, unnaturally cold. All over Europe and North America, snow is falling in June, crops are freezing in the field, and starvation is likely. In fact, this will come to be known as the summer of "eighteen hundred and froze to death."

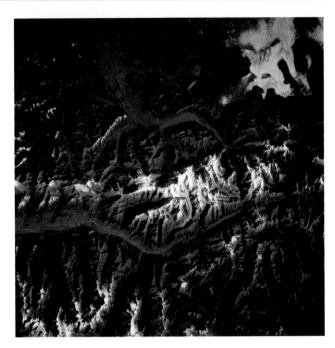

WHILE PERRAUDIN HAD TO PIECE TOGETHER OBSERVATIONS OF BOULDERS AND GLACIERS TO UNDERSTAND THE LANDSCAPE, TODAY WE HAVE ACCESS TO HIGH-TECH INFORMATION. THIS IMAGE OF THE SWISS ALPS TAKEN FROM A SPACE SHUTTLE SHOWS DARK PEAKS ALTERNATING WITH VALLEYS CARVED OUT BY GLACIERS.

PHOTO: NASA Goddard Space Flight Center

The world was in an uproar and people were wondering, Will it happen again? Are summers colder now? A Swiss scientific society, anxious to boost Switzerland's reputation in science, offered a prize to anyone who could contribute to explaining the change in weather.

A man named Jean-Pierre Perraudin of Switzerland wasn't a scientist, but he was an observant man and mountain guide of keen intelligence. Hiking through the Alps, he'd seen huge boulders sitting alone in valleys, as though a giant had walked through and knocked them off the mountains above.

Perraudin noticed that the boulders had long scars in them, which reminded him of scars in rocks far away, near glaciers. This evidence,

combined with his knowledge that glaciers melted or grew sometimes, helped him to take a leap in his thinking. He proposed that glaciers had moved through the Alps, carrying boulders off the mountain crags as they went, and leaving scratches on the boulders.

Eventually, people came to understand that glaciers do move, accompanied by drastic changes in global climate. While this was groundbreaking, it did not explain why the summer of 1816 was so cold. Benjamin Franklin had thought of the answer fifty years before. Volcanic eruptions could cast a pall over the skies worldwide and block so much sunlight that global temperatures dropped. In 1815, an Indonesian volcano had erupted violently, dumping millions of tons of dust into the atmosphere.

Since then, the scientific community has sought answers and identified four important factors in climate change. All of them are rooted in a single idea: climate change is a matter of energy in versus energy out. The earth gets most of its energy from one place, the sun. If the earth takes in more energy than it sends back out to space, the earth warms. If it sends out more than it takes in, the earth cools.

VOLCANIC ERUPTIONS: REFLECTING OR TRAPPING SOLAR RADIATION?

While Benjamin Franklin did correctly surmise that volcanic eruptions can cool the earth, he didn't quite get the whole picture. Volcanic eruptions both warm and cool the earth, but their cooling effect is stronger than their warming effect.

The warming comes from the release of greenhouse gases into the atmosphere. Carbon dioxide, water vapor, and methane are released during volcanic eruptions, and they trap heat that radiates from the earth's surface. The cooling, on the other hand, comes from the tons of rock and ash particles that volcanoes spew into the air—and from liquid droplets and gases. In particular, sulfuric acid in the atmosphere

READING SELECTION

EXTENDING YOUR KNOWLEDGE

reflects and scatters the sun's radiation, acting like a shield against solar radiation and bouncing some of it back into space.

When volcanoes blast sulfur-rich gases into the air, the sulfur reacts with water vapor, and the result is clouds of sulfuric acid. This is why the 1982 eruption of Mexican volcano El Chichon had a much stronger effect on global temperatures than the enormous explosion of Mt. St. Helens in 1980. El Chichon, though it blew less rock and debris into the air than Mt. St. Helens, pumped out more sulfur-rich gases. This caused global temperatures to fall up to five times as much as they did after the Mt. St. Helens explosion.

It's important to understand that volcanic eruptions happen all the time on earth. Whether or not they affect climate has to do with how extreme the eruptions are. After very large eruptions—like Tambora, the Indonesian eruption of 1815, or Krakatoa, a volcanic island which exploded in 1883—the earth may cool by a degree or more for over a year.

VARIATIONS IN THE EARTH'S ORBITAL PATTERNS: THE PLANET COMES CLOSE TO THE FIRE AND WARMS ITSELF

Come near the fire, and you feel how suddenly it warms you; move away, and feel how fast you get cold. Since our main source of energy is the sun, another way for the earth to change temperature is to change how far it is from the sun.

This was the theory developed by a Serbian astrophysicist, Milutin Milankovitch. He noted that the earth changes its distance from the sun in three ways. First, the earth's orbit is not a perfect circle, so it is not always at the same

▸ **VOLCANIC HAZE IN ICELAND FROM THE RELEASE OF GASES**

PHOTO: Andrea Schaffer/creativecommons.org

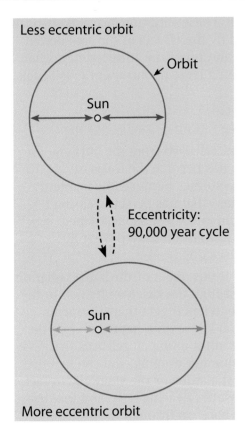

Less eccentric orbit

Orbit

Sun

Eccentricity:
90,000 year cycle

Sun

More eccentric orbit

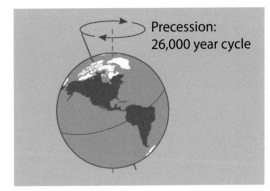

24.5° Line perpendicular
to orbital plane

Axis of
rotation

22.1°

Obliquity:
40,000 year cycle

Precession:
26,000 year cycle

▶ THE TILT OF EARTH
ON ITS AXIS,
OR OBLIQUITY,
CHANGES OVER A
40,000 YEAR CYCLE.

▶ PRECESSION, OR THE
CHANGING ORIENTATION
OF EARTH'S AXIS OF
ROTATION, OCCURS OVER
A 26,000 YEAR CYCLE.

▶ THE SHAPE OF EARTH'S ORBIT, OR
ECCENTRICITY, CHANGES OVER A 90,000
YEAR CYCLE. THE ORBITS IN THE DIAGRAM
ARE NOT DRAWN TO SCALE.

distance from the sun. At its farthest position from the sun, it's about 5 million kilometers (about 3 million miles) further away than at its nearest. Furthermore, the shape of the orbit (or eccentricity) changes. Over a span of about 90,000 years, the earth's orbit goes from being nearly circular to slightly oval and back again The more elliptical, or oval, its orbit, the greater the difference between this maximum and minimum distance from the sun.

Second, the tilt of the earth on its axis (or obliquity) changes, swinging back and forth in a 40,000-year cycle, like a sunbather sitting up in a lounge chair and then adjusting to lean back. The change in the tilt isn't huge, but it's enough to change how directly the southern and northern latitudes face the sun, and that

changes the amount of solar radiation they receive. The more directly we face the sun, the more solar radiation we get.

Finally, as the earth rotates on its axis, it wobbles like a gyroscope as it spins. It does this very slowly, in a 26,000 year cycle. Without this precession, you'd expect that we get closest to the sun in the same season every year. However, because precession makes the earth gyrate, sometimes the planet approaches the sun when it's winter in the Northern Hemisphere, and sometimes it comes near when it's winter in the south. Right now, the earth comes nearest the sun during our winter, which makes winters relatively mild. But 11,000 years ago, the earth came near the sun in July, making northern summers very hot and northern winters very cold.

READING SELECTION
EXTENDING YOUR KNOWLEDGE

By charting the three types of earth motion, Milankovitch calculated the changes in solar energy reaching the earth over the last 600,000 years and matched them to changes in the earth's climate. For about 50 years his theory was ignored, but in the 1970s it was rediscovered by the climatology community, and embraced. When the earth's motion around the sun gives us severe winters in the north, ice and snow build up on the landmasses and last through the summers, and we have an ice age. We're due for another ice age in 50,000–100,000 years.

SOLAR OUTPUT: THE FLICKERING SUN

Given that our main source of energy is the sun, might the planet be warming right now simply because the sun is burning brighter? Astronomers since Galileo have noted the existence of sunspots: dark, cool spots on the sun. These cooler spots (still 3000-4000°C, or 5400-7200°F) are created by huge magnetic storms on the Sun's surface, and they appear and vanish in cycles about 11 years long. During times of high sunspot activity, very bright spots called faculae also appear. Because the effect of the hotter faculae is stronger than the effect of the cooler sunspots, high sunspot activity means increased solar brightness and more energy reaching the earth.

Beginning in the 1980s, satellite data made it clear that sunspots are not the only cause of change in solar energy output. The sun has a lively internal life that affects how it shines. Overall, the sun's energy output does vary by as much as 0.1% over spans of a year or so, and it turns out that this flickering of the sun can cause real changes in the earth's climate. Climatologists consider changes in solar brightness to be a major cause of the last Ice Age.

How much of climate change today is being caused by variations in the earth's orbital patterns and solar flickering? Scientists are still debating this question, but they generally agree that the earth's motion relative to the sun and solar flickering are small factors in the current warming trend. These variations have likely been overshadowed by the effects of the coal-driven Industrial Revolution of the 18th century, when we began burning large quantities of fossil fuels.

CHANGES IN GREENHOUSE GAS PRODUCTION: KEEPING THE HEAT IN

Carbon dioxide, released from the burning of fossil fuels, has become the focus of climate change studies because it is a potent "greenhouse gas" that has been increasing in concentration in the atmosphere since the Industrial Revolution began. What do greenhouse gases do?

In simplest terms, most of the sun's radiation shines right through the gases on its way to the earth's surface. When it gets there, it warms the surface. You know what happens when the ground warms: after a time, it releases that energy. On a summer evening, you can feel the heat coming off a stone or the sidewalk. In fact, the earth is always releasing heat, and that heat energy travels back up through the atmosphere. But when it hits the greenhouse gases,

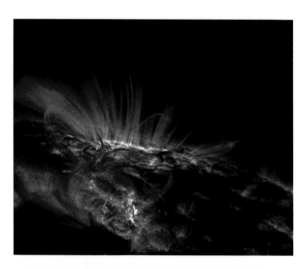

▶ EVEN A RELATIVELY QUIET DAY ON THE SUN IS BUSY. THIS ULTRAVIOLET IMAGE SHOWS BRIGHT, GLOWING ARCS OF GAS FLOWING AROUND SUNSPOTS.

PHOTO: NASA Goddard Space Flight Center

they absorb the energy, trapping it. Slowly, greenhouse gases let the energy back out, but in all directions. Some leaves the atmosphere and goes back into space, and some stays in the atmosphere, warming it.

You might wonder why the gases don't stop the energy as it comes in from space, but do stop it when it's on its way back out. The answer lies in the kinds of waves that carry the energy. The wavelengths that come from the earth's surface are much longer, and much less energetic, than those coming from the sun. As it happens, greenhouse gases are very good at absorbing those longer waves, holding onto their energy, and then transferring it to other molecules in the air as they bump around, making the whole atmosphere more energetic, or warmer. They hardly pick up the short waves at all, though, which means sunlight travels freely through a layer of greenhouse gases.

Greenhouse gases aren't all bad. Without them, we'd be in trouble. Lots of energy would escape, and the surface temperature of the earth would drop by about 14°C (25°F), on average. Too much greenhouse gas, however, functions like a heavy winter coat. Over the past three centuries the amount of one greenhouse gas, carbon dioxide, in the atmosphere has been increasing—and quickly.

Nearly all climate scientists agree now that the main cause of current global warming is our release of greenhouse gases into the atmosphere by burning fossil fuels. Because we seem to be causing the warming, we say that it is anthropogenic. (Anthro- means man; -genic means coming from.)

IT'S COMPLICATED!

All of these factors affect earth's climates. But figuring out how strongly they influence climate, how they affect each other, and even whether they are warming or cooling the earth is an incredibly complex job. We've come a long way from relying just on scars on boulders, but we're still trying to understand an entire planet's climate. ■

DISCUSSION QUESTIONS

1. Why do you think scientists research factors in climate change that may not have much to do with current global warming?

2. What impact do you think a reduction in the greenhouse gases produced by burning fossil fuels might have on Earth and its environment?

CLIMATE AND ENERGY USE

▶ HERE, A SPECIALIST IS INSTALLING A SOLAR TREE, WHICH WILL ROTATE TO FACE THE SUN THROUGHOUT THE DAY AND GENERATE ENOUGH ENERGY TO POWER THREE TO FOUR HOMES. HOW DO ENERGY USE AND GLOBAL CLIMATE CHANGE RELATE?

PHOTO: U.S. Navy photo by Boatswain's Mate 1st Class Christopher Dallaglio

INTRODUCTION

You and your classmates have studied multiple graphs and examined data on changes in the concentration of carbon dioxide in the atmosphere, surface sea temperatures, glacial mass, spring snow cover, arctic sea ice extent, permafrost thawing, land and ocean surface temperatures, and other variables related to climate change. You've also studied climate projections, in which scientists look for trends in climate data, and try to estimate how those trends will translate to everyday climate realities a hundred years from now.

In this lesson you will model ice melt. In your investigation, you will make a prediction based on careful classroom observations and what you know about ice. Then, you'll test that prediction. You will also investigate the effect that global warming may have on our future and ways that impact can be reduced or mitigated.

OBJECTIVES FOR THIS LESSON

▶ Read about scientists' development and use of computer climate models.

▶ Acquire data from an event and use it to estimate the course of a similar event.

▶ Evaluate home energy use.

▶ Consider ways to reduce energy use.

▶ Review the concepts and skills developed in *Understanding Weather and Climate*.

▶ **MATERIALS FOR LESSON 12**

For you

1	copy of Inquiry Master 12.1: Graph Paper
1	copy of Student Sheet 12.1: Predicting the Rate of Ice Melt
1	copy of Student Sheet 12.2a: Personal Emissions Calculator—Student Grid 1
1	copy of Student Sheet 12.2b: Personal Emissions Calculator—Student Grid 2
1	copy of Student Sheet 12: Weather and Climate Review
1	Global Warming Wheel packet
1	paper fastener
1	pair of scissors
3	colored pencils, each a different color

For each group

1	glue stick

For the class

1	beaker, 250 mL
4	ice cubes
1	rubber band
1	square of mesh fabric
1	stopwatch or watch with second hand
1	hair dryer

GETTING STARTED

1 Read "Consensus on Global Warming and Future Climate Models," on pages 217-221. In your science notebook, record your answers to the following questions:

A. In the 1970s, what kind of climate change did scientists warn of, and why?

B. Is it possible that a warming earth could also be cooled by pollution? Explain.

C. What is the IPCC, and what does it do?

D. Are humans the only cause of increasing carbon dioxide concentrations in the atmosphere? Explain.

E. In the 1990s, how did scientists figure out how much of the carbon dioxide concentration increase could be traced to human activities?

2 Discuss your answers to the questions in Step 2 with your classmates.

3 Read "What's the Climate Forecast?" and record your answers to the following questions:

A. How well do Dr. O'Lenic's answers relate to the climate projections and graphs you have seen in the last two lessons? Are the conclusions you draw from them similar to the ones that he has developed in his study of climate?

B. What are some questions about climate forecasting that you would like to ask Dr. O'Lenic?

C. How do you think Dr. O'Lenic's outdoor activities have influenced his views and appreciation of the natural world?

▶ **HOW DO YOU THINK MELTING ICE WILL IMPACT THE POLAR BEAR AND ITS HABITAT?**
PHOTO: Elizabeth Labunski/U.S. Fish and Wildlife Service

WHAT'S THE CLIMATE FORECAST?

Dr. Ed O'Lenic, the senior meteorologist and chief of the Operations Branch of NOAA's Climate Prediction Center, has spent his entire career in short- and long-term weather and climate predictions. He became interested in weather and climate because he saw science "as a way for ordinary, but motivated, people to understand, and do something for the world." He develops and tests complex models of climate change and uses them for short- and long-range climate change predictions. Outside work, the scientist enjoys hiking, and has traveled the Appalachian Trail and other hiking paths around the world.

Recently Dr. O'Lenic answered questions posed to him by a science teacher on behalf of her students. The interview went as follows:

Question: What are the most important things for students to know about global climate change?

Answer: Earth's climate is changing extremely rapidly, mainly through the actions of humans. Many of these changes are permanent; consequences will be large, widespread, and uncontrollable. Action to mitigate, or reduce the effects of, change is needed to eliminate further changes. Adaptation will be necessary and costly.

Question: What do you think is the most compelling evidence for climate change?

Answer: Evidence is everywhere we look, but the long-term trends in temperature are the most compelling. Then, there is the loss of glaciers, decline in arctic sea ice, rising sea level, high nighttime low temperatures, northward migration of tropical diseases, rising acidity of the oceans, [and] extremely high rate of species extinctions.

Question: What do you see as the greatest impact of climate change?

Answer: Modern humans don't seem able to grasp the magnitude of the changes we are experiencing, so they don't have much urgency about them. Our lack of preparation for this, and other similar catastrophes, whose occurrence has uncertainty associated with them, e.g. [a] large volcanic eruption, meteor impact, or solar storm, leaves humanity at a survival disadvantage.

Question: What are steps that students and a school community can take to make a difference in the energy we use and the carbon dioxide we produce?

Answer: The most accessible measure is improving the efficiency of heating and cooling our homes and offices by improving insulation and by lifestyle changes, such as raising the thermostat a degree in summer and lowering it a degree in winter.

Question: When did the National Weather Service and other governmental agencies start to investigate data indicating climate change?

Answer: Svante Arrhenius, in 1896, was the first scientist to correctly discuss in public the implications of the changes of carbon dioxide concentration in the atmosphere.

Question: What is your response to the skeptics of global change?

Answer: Nature cannot be fooled.

Dr. O'Lenic's response to the last question was based on his belief that nature or the earth is a system of inter-linked parts that have predictable relationships, regardless of human doubt or lack of understanding. ■

INQUIRY **12.1**

MODELING THE RATE OF ICE MELT

PROCEDURE

1 You have learned about what can affect our planet's climate, and you have explored climate projections from scientific models. In this inquiry, with your teacher, your class will build a model to explore the rate of ice melt. Listen as your teacher explains the experiment.

2 Set up a table in which you can record data about the melt rate of an ice cube. In your science notebook, make a T-chart with two columns: "Time" and "Volume." 🖎

3 Listen to your teacher's instructions for conducting the class experiment. Your teacher will ask for student volunteers to help keep time and measure the amount of water collected in the beaker. Record the measurements taken during the experiment in the table you created.

4 Use the data you recorded and a colored pencil to make a graph on Inquiry Master 12.1: Graph Paper that shows the melt rate of one ice cube. Label one axis "Time" and the other "Volume." Make the axes long enough to accommodate 10 minutes and 30 mm. Label the line you plot "One Ice Cube."

5 Look carefully at your graph. It can help you predict the rate of melting for three ice cubes. With your group, discuss the rate at which you think three ice cubes will melt. You've had lots of experiences with melting ice. Recall the experiment you just conducted. Picture a few cubes melting in the bottom of a glass. Consider what factors will affect the way they melt, and how long they will take to melt. Think about how much water will collect in the beaker every 30 seconds. Plot these points on your graph on Inquiry Master 12.1 using a second colored pencil. Connect these points using a dashed line and label the line "Three Ice Cubes (Prediction)." In your science notebook, record your reasoning for why the ice cubes will melt this way. 🖎

6 Now the class will test its predictions. Draw another table for recording time and volume data.

7 Work with the class to conduct the experiment a second time, using three ice cubes. Record the data collected during the experiment.

8 When all three ice cubes have melted completely, use a third colored pencil to plot the data points on your graph on Inquiry Master 12.1 and draw the line they suggest. Label this line "Three Ice Cubes (Actual)."

9 Answer the questions on Student Sheet 12.1. Then, discuss your ideas with the class.

A. On your graph of the rates of ice melt, is the line showing how fast the three ice cubes actually melted the same as the dashed line you drew for your prediction? If not, how does it differ?

B. What factors do you think might account for differences between these lines?

C. When you made your prediction and drew the dashed line, what assumptions did you make about how ice melts?

D. Could you use this process to predict how long an iceberg might take to melt? Why or why not? What else would you want to know before you tried to estimate an iceberg's melt time?

CALCULATING YOUR CARBON FOOTPRINT

PROCEDURE

1 Read the introduction, "What Is a Carbon Footprint?" on Student Sheet 12.2a: Personal Emissions Calculator—Student Grid 1. The information on this sheet will help you understand the carbon footprint each of us leaves on the earth and ways to reduce that impact on the environment.

2 This activity on energy emissions and savings was developed by the United States Environmental Protection Agency. To get started, follow these directions to create a global warming wheel, which is needed for the inquiry:

A. Cut out the two wheels and the rectangular pieces in the Global Warming Wheel packet.

B. Align the wheels with the blank sides together and match up the four labels on the outside of the circles (Waste Disposal, Electricity Use, Home Heating, Transportation).

C. Cut out the large and small rectangular windows from each of the rectangular sheets "What's Your Score?" and "What Can You Do?" from the Global Warming Wheel packet.

D. Align the small circle on the wheel of energy use with the small circle on the "What's Your Score?" sheet. The words on the wheel will appear in the windows on the sheet. Place the paper fastener though the two small circles.

Inquiry 12.2 continued

E. Turn the sheet and wheel combination over and place the second sheet, "What Can You Do?" so that the small circles line up and the smaller window is closer to you. Again, the words on the wheel will show up in the windows of the sheet. Finish connecting the sheets and the wheel with the paper fastener.

F. Glue the corners of the sheets together so that the wheel turns and the pieces hold together.

G. You can highlight the edges of the rectangular windows to make it easier to read.

3 If you haven't done so already, complete Student Grid 1 on Student Sheet 12.2a.

4 Read the Introduction on Student Sheet 12.2b: Personal Emissions Calculator—Student Grid 2. Use the "What's Your Score?" side of the global warming wheel to calculate your emissions per year. Record this information on your student sheet.

5 Work with your classmates to calculate the class average for emissions, and record this at the bottom of Student Sheet 12.2b.

6 Use the "What Can You Do?" side of the global warming wheel to calculate potential emissions reductions for your family. Record this information on Student Sheet 12.2b.

7 Read "Alternatives to Fossil Fuels," on pages 222-227.

REFLECTING
ON WHAT
YOU'VE DONE

1 Discuss these questions with your group and write your answers in your science notebook:

A. What did Inquiry 12.1 show you about the kinds of problems scientists think about when they try to make models that predict how climate change will affect some part of the earth?

B. What can you do as an individual to reduce your energy consumption and your carbon footprint?

C. What can you do as a class or a school to reduce your energy consumption and carbon footprint?

Consensus on Warming and Future Climate Predictions

Scientists have collected evidence of global climate change for decades. What the data mean, though, has been the subject of intense debate over the years.

In the 1970s, many scientists forecast a new ice age, with the planet cooling to a point not seen since glaciers covered the upper Midwestern United States. The forecast was based on research on air pollution coupled with a brief cooling trend. At the time, the skies were badly polluted with industrial smog, soot, and exhaust. The pollution made for hazy skies and dim sunlight, and scientific evidence showed that an atmospheric blanket of soot—manmade or volcanic—could block enough sunlight to lower the earth's temperature.

But atmospheric scientists at this time were also aware of a "greenhouse effect," the idea that certain gases in the atmosphere could trap heat radiated and reflected from the earth's surface. It was thought that greenhouse gases might actually protect the planet from a new ice age. Even so, by the 1980s some scientists were sounding an alarm. Not only had they found levels of greenhouse gases rising quickly, but it now looked as though the earth's temperature was rising, not falling. In the ensuing confusion, scientists met regularly to discuss whether or not global warming was happening. If it was happening, how fast was the planet warming, and just how warm might it get? What might that mean for life on earth? And why was it

happening? There was a great deal they'd have to learn about earth's climate, they realized, before they could begin to answer these questions.

In 1988, the United Nations formed an international group of highly esteemed scientists to examine the climate-change science being published—thousands of papers—and report on what they read. This was the Intergovernmental Panel on Climate Change (IPCC), and it is still recognized as a major authority on climate-change science.

The IPCC determined that the planet was indeed warming. At first the findings were tentative, but evidence increased as researchers asked new questions related to global warming, and collected data to answer them. There was disturbing evidence that the planet might be warming faster than scientists had originally thought—and there were dismaying reports from polar scientists, who spoke urgently about rapid glacier melting, ice shelf disintegration, and permafrost melting. Consensus that the earth was warming strengthened—and the IPCC's second report said humans were likely at fault.

Attention turned, urgently, to *why* the earth was warming. Was it really because of greenhouse gases humans had released into the atmosphere? Was it a natural process? Or was it a combination, and if so, how much were humans responsible for? These were important questions: if global warming was mainly a result of greenhouse-gas pollution, we would have to

SCIENTISTS DRILL HOLES IN TAYLOR GLACIER IN ANTARCTICA TO RECORD ITS TEMPERATURE, MOVEMENT, AND MELTING. THIS INFORMATION HELPS IN MONITORING GLOBAL CLIMATE CHANGE.

PHOTO: Kristan Hutchison/National Science Foundation

change our way of life—and do it fast. It would be an expensive, difficult change, and not one to be undertaken unless it really was necessary.

These were also tough *scientific* questions to answer. Humans are not the only source of greenhouse-gas pollution on earth. The earth itself releases gases into the atmosphere all the time through volcanoes and other vents. How could human-generated, or anthropogenic, pollution be separated from the earth's?

In the 1990s, the only way to answer the question was to measure total atmospheric carbon dioxide levels, find the change from the year before, and then try to figure out how much carbon dioxide humans had added to the atmosphere. Scientists did that by totaling the amount of fossil fuel sold worldwide in the previous year, assuming it had all been burned, and calculating how much carbon dioxide it would have released. The difference between the total carbon dioxide change and our contribution would show the percentage for which humans were responsible. This was only a rough estimate, and the imprecision gave scientists and policymakers plenty to argue about.

The IPCC continued to produce reports. In 2007, its fourth report said that, taken together, the world's climate studies indicated a 90–99% chance that global warming was anthropogenic, and that the earth had warmed by 0.75°C in the last hundred years. It was real, it was still going on, and it was almost certainly our doing.

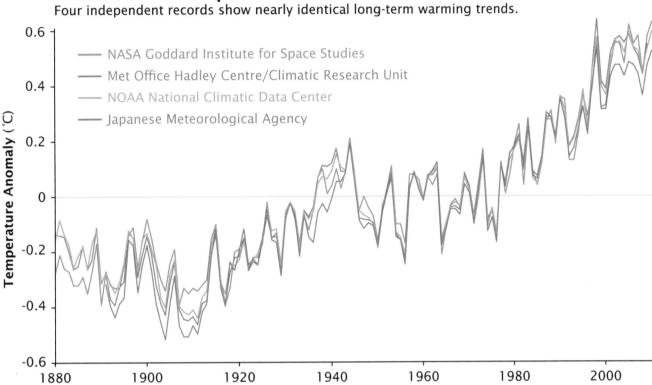

Global Surface Temperatures
Four independent records show nearly identical long-term warming trends.

Legend:
- NASA Goddard Institute for Space Studies
- Met Office Hadley Centre/Climatic Research Unit
- NOAA National Climatic Data Center
- Japanese Meteorological Agency

Y-axis: Temperature Anomaly (°C)
X-axis: Year

▶ SCIENTISTS FROM SEVERAL MAJOR INSTITUTIONS—NASA'S GODDARD INSTITUTE FOR SPACE STUDIES (GISS), NOAA'S NATIONAL CLIMATIC DATA CENTER (NCDC), THE JAPANESE METEOROLOGICAL AGENCY, AND THE MET OFFICE HADLEY CENTRE IN THE UNITED KINGDOM—TALLY THE TEMPERATURE DATA COLLECTED AT STATIONS AROUND THE WORLD AND MAKE INDEPENDENT JUDGMENTS ABOUT WHETHER THE YEAR WAS WARM OR COOL COMPARED TO PREVIOUS YEARS. ALL FOUR RECORDS IN THIS GRAPH SHOW PEAKS AND VALLEYS IN SYNC WITH EACH OTHER. ALL SHOW PARTICULARLY RAPID WARMING IN THE PAST FEW DECADES. AND ALL SHOW THAT THE LAST DECADE WAS THE WARMEST.

PHOTO: NASA Earth Observatory/Robert Simmon, Data Sources: NASA Goddard Institute for Space Studies, NOAA National Climatic Data Center, Met Office Hadley Centre/Climatic Research Unit, and the Japanese Meteorological Agency.

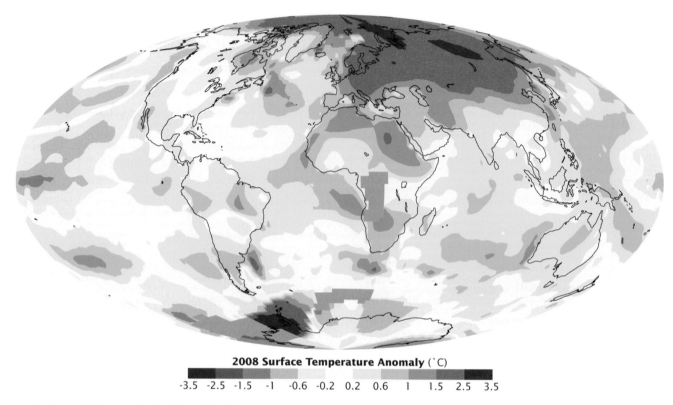

2008 Surface Temperature Anomaly (˚C)

-3.5 -2.5 -1.5 -1 -0.6 -0.2 0.2 0.6 1 1.5 2.5 3.5

▶ THIS MAP SHOWS GLOBAL TEMPERATURES IN 2008 IN RELATION TO AVERAGES FROM A BASELINE PERIOD FROM 1950–1980. BELOW-AVERAGE TEMPERATURES (I.E., LESS THAN THE BASELINE PERIOD) IN 2008 ARE SHOWN IN BLUE, AVERAGE TEMPERATURES ARE WHITE, AND ABOVE-AVERAGE TEMPERATURES ARE RED.

PHOTO: NASA image by Robert Simmon, based on GISS surface temperature analysis data

CLIMATE MODELING AND PREDICTION

So what does all the IPCC's information mean for us? Does it really matter if the earth warms by a couple of degrees? If we stop driving cars, does global warming stop? To answer these questions, scientists have taken data on how the climate has changed before—over hundreds, thousands, millions, and billions of years— and tried to determine how and why climates changed in the past. What happened to the earth when it last warmed, when glaciers melted and their weight was no longer on land masses? What happened to the weather? What happened to life when the earth warmed slowly, and what happened when it warmed quickly?

Answering these seemingly simple questions is a very complex process. It requires trying to determine the relationships between thousands of measurable changes. For instance, for every tenth of a degree that the surface atmosphere warms, how far northward do tropical plants move?

Building a reliable description of these relationships is called building a model, or an idea of how things work. Models are built on assumptions. For instance, scientists building a hurricane model might assume that an increase of 2 percent in sea surface temperature (SST) increases the hurricane's rainfall by 8 percent. Another team might assume that a 2 percent increase in SST increases hurricane rainfall by 14 percent.

Why do different groups of scientists make different assumptions? Because the evidence varies. If there is real evidence suggesting that a 2 percent increase in SST bumps up rainfall by 8 percent sometimes, and 14 percent sometimes, either assumption might be valid. As scientists study hurricanes more and come to understand them better, they may be able to say, "Well, yes, rainfall goes up by 14 percent sometimes, but only when a few other things also happen." Then they can refine the models to reflect their deeper understanding of hurricanes. They might also understand better when to use each model, choosing among them as if choosing the right tool from a toolbox.

By asking precise questions about climate, building models, and combining the models, scientists can predict with increasing accuracy and precision how the earth might change as it warms.

A problem with using models to predict the future is that the future hasn't happened yet, so scientists can't test whether a model is correct. What they can do, though, is see how well models work on the past. If a model is applied backwards in time, do the results match the actual past? This type of model testing is called hind-casting, or back-testing.

Our knowledge about the earth's environment has grown explosively since scientists talked about a new ice age. Urgent questions have driven its growth. But as you explore climate change modeling, take time to enjoy the great payoffs on the way to the answers: the beauty of the models, and just how amazing it is to see and know so much about the earth as a dynamic planet. ■

DISCUSSION QUESTIONS

1. What is hindcasting and how does it help us understand climate change?

2. In the 1970s, many scientists thought the earth was cooling. Now most believe the earth is warming. Should we trust them to make predictions? Why or why not?

Alternatives to Fossil Fuels

in the context of the long history of carbon dioxide concentrations in the atmosphere, current levels are relatively low. However, they are rising rapidly, and because carbon dioxide is a greenhouse gas, its increase contributes to global warming. Reducing global warming is one incentive for a search for renewable alternatives to fossil fuels, which release carbon dioxide when they're burnt.

Many alternate sources of energy have been around for millennia—thousands of years before we began mining coal, oil, and natural gas from the earth. Wood has been used as a fuel since fire was discovered. The Greeks and Romans used the power of falling water to turn wheels. These ancient sources, along with newer ones, offer promise for renewable sources of power.

Other alternative, renewable sources of energy take advantage of modern chemistry and physics, and have been in use for only a few decades. What all non-fossil energy sources have in common, though, is that they come with their own costs and problems. Here, we'll look at the pros and cons of several alternatives to fossil fuels.

WIND POWER

Windmill blades are attached to a generator housed at the top of a 60- to 90-meter (200- to 300-foot) pole. When wind blows, windmill blades turn a shaft, which spins a turbine even faster. Inside the turbine is a shaft wrapped in wire cables; around the wire lies a ring of magnets. When the wire spins inside this magnetic field, an electric current is generated in the wire.

If you've seen a wind farm at work, it might look as though the blades are turning slowly. In fact, because the blades are so long—about 21 to 40 meters (70 to 130 feet)—they're turning very fast at the tip. The blade tips can reach speeds of over 90 meters (295 feet) per second in a good wind.

The use of wind power has grown from small windmills on individual farms to large numbers arrayed in wind farms. Over the last decade, the amount of electricity generated by wind power has increased by 25 percent per year. Despite the increase, this amount is still small; by the end of 2010, only 2.5 percent of the world's electricity came from wind power.

▶ **WIND FARM IN COPENHAGEN, DENMARK**

PHOTO: www.cgpgrey.com/creativecommons.org

Windmills have the advantage of working in any place where the wind blows strongly and consistently. They work well in the ocean and in many parts of the United States. They produce no carbon emissions, smog, or smoke, and wind power is renewable as long as the sun shines.

So what's the downside? The amount of power produced by wind farms changes with the wind, and they often have to be placed far from population centers. Although they're clean, they're not pollution free: they produce noise and light-effects pollution. The light flicker they cause can make people feel ill, and the shadows cast by the blades and windmill towers can damage property values. They can also interfere with bird migration patterns and weather radar.

There's also the question of what effect taking all that energy out of the atmosphere might have. The atmosphere has a "downstream" just as rivers do, and we don't know yet what weaker winds downstream might mean for climate. Early evidence says that wind farms have no perceptible downstream effect, but we need to know what problems might emerge if we tried to get a large part of our energy from wind.

▶ **SOLAR ARRAY**

PHOTO: Photo by Jurvetson (flickr)

SOLAR POWER

When we talk about using solar power to generate electricity, though, we are talking about capturing sunlight's energy ourselves and converting it directly into electrical or heat energy. This energy can be captured by using photovoltaic cells, which are sensitive to individual photons, or particles of light.

In the top layer of a photovoltaic cell are atoms whose electrons can be knocked out of position by photons. These electrons can be made to flow in order to generate an electrical current.

Since many solar cells need bright sunlight to work well, their energy output varies with weather and the seasons. People who use solar power systems store energy on bright days, and can use backup sources of electricity on cloudy days and at night. The cost of photovoltaic-system installation and maintenance remains relatively high—it's unlikely, for instance, that your home will have solar cells on the roof—but the costs have come down in the last decade, and new materials are making solar cells more efficient. It's possible that someday photovoltaic cells will produce energy as cheaply as fossil fuels do today, even in places that don't get lots of bright, direct sunlight.

GEOTHERMAL ENERGY

Geothermal energy comes from heat generated deep within the earth by radioactive atoms. This heat is conducted from the molten core through the bedrock to the surface. Hot springs, deep-water reservoirs, and steam vents are all sources of geothermal heat.

Some people heat their homes by pumping water dozens of feet into the ground, where it's warmed and returned to their homes. But power generation plants need hotter water or even steam to drive turbines. They must go deeper into the earth—sometimes miles deep—or use heat that comes from magma welling up through the crust. Recently, some companies have begun cracking bedrock deep in the earth, pumping water down to it, and returning heated steam to the surface.

Currently less than 1% of the United States' electricity is geothermally generated, though in a few parts of the world, 10-25% of electricity comes from the earth's heat. Geothermal is renewable, though individual wells may eventually run out of water or cool off. There are risks with geothermal energy. Geothermally heated steam can carry dissolved toxic elements and compounds from deep within the earth to the surface. They can damage the environment if they're released. Also, cracking bedrock for geothermal power production can cause earthquakes. This cracked bedrock can't be repaired, if in the future we discover that this has lead to unintended problems.

BIOFUELS

Plants and animals are biomass: the living weight of the world. Wood is a biomass fuel, or biofuel. In many countries, people burn plant and agricultural waste such as the stubble left in fields after harvest. Biofuel also comes in the form of methane, a gas released by bacteria in landfills as they break down plant and animal wastes. Certain crops, such as corn, sugar cane, and switchgrass, can be converted to alcohols and burned as motor fuel. This, too, is biofuel. Biodiesel is a biomass replacement for gasoline, a fossil fuel made from petroleum. Biodiesel is made from vegetable oil, used cooking grease from restaurants, and animal fat.

Biofuels do not, on balance, add carbon dioxide to the atmosphere. Organisms take carbon from the air while living, and return it to the air when burnt. Unfortunately, producing biofuels adds carbon dioxide to the atmosphere. Cheap diesel and gasoline power the equipment used to grow, harvest, process, and transport biofuels.

Clearing land so that farmers can grow biofuels also causes problems. Grassy fields and forests are "carbon sinks:" places where carbon is taken out of the atmosphere and kept out for a long time. Killing those plants and disrupting their soil for biofuel farms destroys that carbon sink. In addition, using farmland to grow fuel also reduces the amount of land and water that can be used to grow food crops, and this may lead to food shortages or higher food prices.

Scientists studying the future of biofuels are studying single-celled organisms, like algae, which can be grown in salt water, and won't compete with edible crops. Studying algae is not so easy to do: scientists have been working on it for over half a century. Algae farms are easily contaminated. So far, biologists have found that in order to keep large algal cultures alive and well, they need to grow them in controlled environments, in tanks or artificial ponds. And that takes—energy.

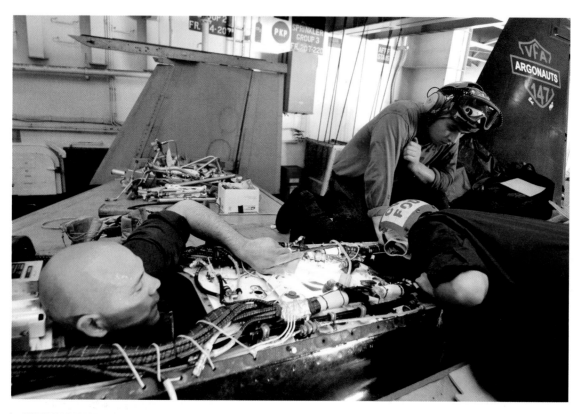

▶ **EXPERTS WORKING TO INSTALL A FUEL CELL INTO AN F/A-18E SUPER HORNET AIRCRAFT.**

PHOTO: U.S. Navy photo by Mass Communication Specialist 3rd Class Shawn J. Stewart

HYDROGEN FUEL CELLS

The hydrogen fuel cell consists of three parts sandwiched together. The "bread slices" are an anode (positively charged) and a cathode (negatively charged); between them is an electrolyte, or liquid, that can carry electrical charges.

The anode is a metal plate, usually platinum, that can strip hydrogen of its electron. All that's left of hydrogen atoms after stripping are the protons which are pulled through the electrolyte to the cathode. The electrons would like to follow, but the electrolyte won't let them pass through. They must get to the cathode by flowing outside the cell and back again to the cathode, and that flow is an electrical current. Once electrons reach the cathode, they and the protons react with oxygen, which is also steamed into the cathode. The reaction creates water (H_2O) and generates heat.

Hydrogen fuel cells can be used to power many things, from cars to electrical plants. There's a problem though. Platinum, the favored anode metal, is very expensive. Over time, substitutes are being developed that may prove to be important in getting fuel cells into widespread use. Another issue is how hydrogen fuel stations can be set up and used safely, since hydrogen is explosive.

NUCLEAR POWER

Nuclear power plants currently produce about 20 percent of the world's energy. Nuclear plants split atomic nuclei in a controlled setting to produce very large quantities of energy, which heats water and turns it to steam. The steam drives turbines, which produce electrical current.

Nuclear power does not produce greenhouse gases, and many environmentalists promote its use. However, splitting atoms does produce dangerously radioactive waste, which remains radioactive for thousands of years. There currently is no satisfactory means of storing this waste. In the past, nuclear waste has been dumped into the ocean; the United States has shipped nuclear waste to a hollowed out mountain in Nevada called Yucca Mountain. In 2011, Congress voted to stop funding the nuclear waste repository there, leaving the United States with no long-term nuclear waste storage facility.

Nuclear power plants can be extremely dangerous if explosions damage the reactors and release radioactive gases to the air, or leak radioactive water to the ground. Heavy exposure to radioactivity has a number of impacts on humans, including cancers and birth defects.

There have been serious accidents at nuclear power plants, releasing radioactivity into the environment. Three Mile Island nuclear plant in Pennsylvania averted a major disaster in 1977, but still released radioactivity into the air. In 1986, an explosion at a nuclear power plant in Chernobyl, Ukraine, contaminated much of Europe and the western Soviet Union with radiation. And in 2011, an earthquake and tsunami severely damaged nuclear reactors at Fukushima, Japan, releasing radioactive material into groundwater and the air.

In the wake of the Fukushima disaster, many countries are having second thoughts about nuclear energy. Because the Fukushima plant leaked radiation after being hit by both an earthquake and a tsunami, many governments are considering how to strengthen nuclear plants so that they can withstand natural disasters.

WHAT'S THE ANSWER?

Do the risks associated with alternative energy sources mean we should just stick to fossil fuels? No! Risks are also associated with the use of fossil fuels—burning them adds carbon dioxide to the atmosphere. Furthermore, we are compelled to explore and use alternatives because fossil fuels will eventually run out. We need to do more research into alternative fuels to discover how we might use them in ways we can live with—more cleanly, more cheaply, and more efficiently. ■

DISCUSSION QUESTIONS

1. Why are we not already reliant just on alternative energy sources, such as wind power, solar power, and geothermal power? Explain what has limited the use of these alternatives.

2. Famous scientists have come out in support of building more nuclear power plants, saying they are a reliable source of power and less dangerous to the environment than burning more fossil fuels. Do you agree? How could you measure the relative dangers and decide in a scientific way?

UNDERSTANDING WEATHER AND CLIMATE ASSESSMENT

▶ **WHAT WILL HAPPEN TO THE HELIX ABOVE A HOT LAMP?**

PHOTO: Terry McCrea/Smithsonian Institution

INTRODUCTION

You have now completed the unit *Understanding Weather and Climate.* You will be assessed on your ability to show what you know and what you can do as it relates to the inquiries you have completed on weather and climate concepts. The assessment in this lesson is divided into two parts. During Part A of the assessment, you will observe a paper helix, or spiral, above a hot lamp and record your observations of the paper helix. You will be asked to describe what the helix tells you about how air moves and about similar movements in the vortex of a storm.

For Part B of the assessment, you will be asked to complete multiple-choice and short-answer questions about weather, forecasts, clouds, tornadoes, hurricanes, ocean currents, climate, and climate changes. You will review diagrams and interpret data. Your teacher will use the results of this assessment to determine how well you can apply the concepts, knowledge, and skills you have learned in this unit. You will also evaluate your own learning by assessing your ability to understand and answer the questions.

OBJECTIVES FOR THIS LESSON

▸ Review and reinforce concepts and skills from *Understanding Weather and Climate*.

▸ Complete a three-part assessment of the concepts and skills addressed in *Understanding Weather and Climate*.

▸ Apply your knowledge and skills to answer questions.

▸ **MATERIALS FOR LESSON 13**

For you

1   completed copy of Student Sheet 12: Weather and Climate Review

1   copy of Inquiry Master 13.1b: *Understanding Weather and Climate* Assessment Questions (Part B)

1   copy of Student Sheet 13.1a: *Understanding Weather and Climate* Performance-Based Assessment Answer Sheet (Part A)

1   copy of Student Sheet 13.1b: *Understanding Weather and Climate* Written Assessment Answer Sheet (Part B)

For your group

1   concept map (from Lesson 1)

For the class

1   brainstorming list, "What We Want to Know About Weather and Climate (from Lesson 1)

GETTING STARTED

PROCEDURE

1 Your teacher will go over the correct responses for the questions on Student Sheet 12: Weather and Climate Review, which you completed for homework.

2 Make sure you ask questions about any topics that aren't clear to you.

PART A: PERFORMANCE-BASED ASSESSMENT

PROCEDURE

1 Obtain one copy of Student Sheet 13.1a: *Understanding Weather and Climate* Performance-Based Assessment (Part A).

2 Listen as your teacher describes Part A and Part B of the assessment.

3 Complete Step 1 on Student Sheet 13.1a. Then watch as your teacher holds the paper helix above the lamp, as shown in Figure 13.1. Record your observations under Step 2 on the student sheet.

4 Complete Student Sheet 13.1a, then turn it in to your teacher.

PART B: WRITTEN ASSESSMENT

PROCEDURE

1 Obtain one copy of Inquiry Master 13.1b: *Understanding Weather and Climate* Assessment Questions (Part B) and one copy of Student Sheet 13.1b: *Understanding Weather and Climate* Assessment Answer Sheet (Part B). Write only on the answer sheet. Inquiry Master 13.1b will be used by other classes throughout the day.

2 Complete the assessment as directed by your teacher.

3 When you have completed the written assessment, turn in Student Sheet 13.1b and Inquiry Master 13.1b to your teacher.

SAFETY TIP

Do not touch the metal reflector on the clamp lamp while it is turned on or is cooling.

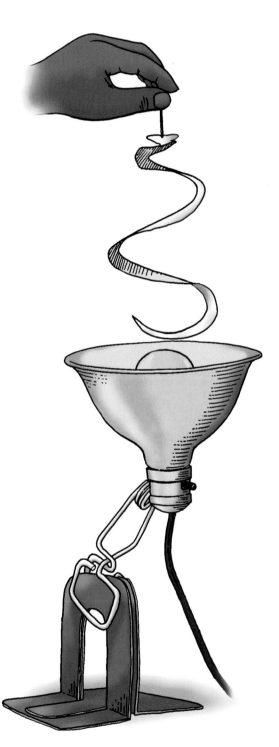

▶ **THE SETUP FOR THE PAPER HELIX.**
FIGURE **13.1**

PART C: REFLECTING ON WHAT YOU'VE DONE

PROCEDURE

1 Review your scored assessment with your teacher. To what extent did your results meet your expectations?

2 Discuss with the class how you can apply what you have learned in *Understanding Weather and Climate* to your daily life. Answer these questions with your group and class:

A. What have you learned about weather that can help you?

B. What was most interesting to you in your study of climate and climate change? Why?

C. What have you learned about how scientists decide which data to collect?

D. What have you learned about how scientists model weather and climate?

E. Climate scientists often worry that they are not clearly communicating the results of their work and its importance to non-scientists. Given what you have learned in this unit, do you agree? Do you think most people know about the things you have learned and investigated? If not, how do you think scientists could better show the public their work and their findings?

3 Look at your group concept map and brainstorming list from Lesson 1. What information do you want to add to the list? Can you answer any of your earlier questions now? What are some new questions that you want to add?

Glossary

abyssopelagic zone: The region of the ocean from 4000 to 6000 meters below the surface. Its name derives from the Greek belief that the ocean had no bottom.

aerosol: Solid particles in the air or in a gas.

air mass: A body of air with the same temperature and moisture throughout.

air pressure: The weight of air.

atmosphere: The thin blanket of gases that surrounds the earth.

bar: A unit of atmospheric pressure.

barometer: An instrument used to detect and measure changes in air pressure.

barometric pressure: The pressure exerted by the atmosphere.

bathymetric map: A two-dimensional map that shows the depth of underwater objects by means of contour lines. See also *contour line; topographic map.*

bathypelagic zone: The ocean's "midnight zone," extending from 1000 to 4000 meters below the surface.

carbon dioxide: A potent greenhouse gas which can be produced by burning fossil fuels.

canyon: A deep gorge in the ocean or on land.

chronometer: A navigational timekeeping instrument which first allowed sailors to measure longitude.

climate: Weather conditions that are characteristic of a region or of a particular place over a long period of time. See also *weather.*

climate model: A predictive representation of a climate, with the climate's phenomena reduced to mathematical expressions.

climate prediction: A short-term (generally less than 90 days) forecast of climatic conditions for a given region.

climate projection: A long-term estimate of how climate might reasonably look over long terms such as decades or a century, given current data and climate models.

cloud: A collection of billions of tiny droplets of water or ice and dust particles visible from the ground.

cold front: The leading edge of a cold air mass. It occurs when a cold air mass pushes a warm air mass ahead of it. See also *weather front.*

condensation: The process by which water vapor changes from a gas to a liquid.

continental shelf: A landform extending from the shore to the steep slope of the ocean basin.

continental slope: A landform that connects the continental shelf to the ocean floor.

contour interval: The space between two adjacent contour lines.

contour line: A line that connects all the points at a particular elevation or depth in an area of a contour map.

contour map: A two-dimensional map that shows elevation or depth by using contour lines. See also *bathymetric map; topographic map.*

convection: The process by which heat moves efficiently through air or water.

convection current: A circulating flow of air or water resulting from temperature differences; also called a convection cell.

Coriolis effect: A force that comes from the rotation of the earth and affects the path of objects in motion.

cumulus cloud: Puffy, rounded clouds that formed at low altitudes.

current: The movement of a gas or liquid in a definite direction. An ocean current is one example.

cyclone: A massive, rotating storm that forms in the Indian Ocean and off the coast of Australia. It is equivalent to a hurricane in the Atlantic Ocean, eastern Pacific Ocean, or Caribbean Sea. See also *hurricane; typhoon.*

deep-sea trench: A deep, narrow depression in the seafloor.

dropsonde: A probe parachuted directly into a hurricane that transmits information which allows meteorologists to determine air pressure, wind speeds, humidity, and temperature inside the hurricane.

drought: An extended period of lack of precipitation.

easterlies: Global winds that flow from the east to the west. See also *trade winds; westerlies.*

El Niño: An unusually warm flow of surface water that occurs in the Pacific Ocean about every three to six years.

Enhanced Fujita Scale: A scale for measuring the strength of tornadoes.

epipelagic zone: The "sunlight zone" of the ocean, ranging from the surface to a depth of 200 meters.

evaporation: The process by which water changes from a liquid to an invisible gas called water vapor.

exosphere: The outer region of the atmosphere.

foraminifera: An order of single-celled marine animals with hard outer shells; also, foraminifers.

geologist: A scientist who studies the history and structure of the earth as it is recorded in rocks.

global conveyor: See *thermohaline circulation.*

global winds: Giant convection currents that circulate within the Northern and Southern Hemispheres of the earth. See also *easterlies; trade winds; westerlies.*

globe: A spherical model of the earth.

greenhouse gases: Gases in the earth's atmosphere, such as water vapor and carbon dioxide, that absorb heat energy radiated from the earth and release some of it earthward, preventing heat's escape into space.

Gulf Stream: A warm-water ocean current that flows north along the East Coast of the United States.

gyre: An ocean-wide circulation of surface (wind-driven) currents.

hadal zone: The deepest parts of the ocean, including everything below 6000 meters.

hurricane: A massive rotating storm with wind speeds of 119 kilometers per hour or more that forms north of the equator in the Atlantic Ocean, eastern Pacific Ocean, or Caribbean Sea when warm air rises over tropical waters. See also *cyclone; typhoon.*

hurricane eye: The calm center of a hurricane.

hurricane eyewall: The wall of rotating clouds surrounding a hurricane's eye. Hurricane winds are strongest in the eyewall.

IPCC: Intergovernmental Panel on Climate Change

Köppen climate classification system: The original and still one of the most widely used climate classification systems.

Kuroshio current: A clockwise Pacific surface current whose equatorial waters warm Japan and the Pacific Northwest coast.

land breeze: The flow of air from land to water, caused by differential heating of sea and land and the different speeds at which they release heat to the atmosphere. See also *sea breeze.*

map: A representation of the earth or a part of the earth, usually on a flat surface.

mesopelagic zone: The ocean's "twilight zone," which extends from 200 to 1000 meters below the surface.

mesosphere: The part of the atmosphere between the troposphere and the thermosphere in which the temperature decreases with altitude.

meteorologist: A scientist who studies the earth's atmosphere and who monitors, studies, and forecasts weather.

methane: A greenhouse gas produced by burning coal and natural gas; also produced by ruminants and landfills.

mid-ocean ridge: A mountain-like landform that develops underwater where tectonic plates separate, magma wells up, and new ocean lithosphere forms.

NASA: National Aeronautic and Space Administration.

natural catastrophic event: A powerful and often dramatic force of nature that changes the earth's surface and atmosphere; includes earthquakes, volcanoes, and intense storms such as hurricanes and tornadoes.

NCDC: National Climate Data Center.

NHC: National Hurricane Center.

nimbus cloud: A tall, dark rain cloud.

nitrous oxide: A greenhouse gas.

NOAA: National Oceanic and Atmospheric Administration.

NWS: National Weather Service.

oceanographer: An older name for marine scientists, which comes from the fact that early oceanographers were primarily concerned with mapping the oceans.

occluded front: A boundary that occurs when both a cold and a cool air mass collide with a warm air mass, which becomes trapped and lifted between them. See also *weather front*.

ozone (O_3): A gas found in the lower level of the troposphere as a component of smog, and in the upper layer of the atmosphere as a barrier against ultraviolet rays.

paleoclimatology: The study of ancient climates, in which scientists attempt to reconstruct ancient climates through the use of proxy data.

proxy data: Data that represents a desired value that cannot be determined directly, e.g., temperatures in climates long ago.

radiation: The process by which energy is transferred from one object, such as the sun, to another object without the space between them being heated.

risk: Exposure to the chance of injury or loss.

Saffir-Simpson scale: A scale used to measure hurricane strength.

salinity: Saltiness.

sea breeze: The flow of air from water to land, caused by differential heating of sea and land and the different speeds at which they release heat to the atmosphere. See also *land breeze.*

seamount: A mountain rising from the ocean floor that does not reach the water's surface.

stationary front: A boundary that occurs when two air masses move close to one another, but neither has enough force to move the other; they both remain fixed in place. See also *weather front.*

storm surge: An unusually high water level, caused primarily by strong winds, especially those associated with a hurricane.

stratosphere: The layer of the earth's atmosphere above the troposphere. It has very little water vapor or other gases; protective ozone layer forms there.

stratus cloud: A low-lying, layered cloud.

temperature: A measure of how hot or cold a material is; an indication of the amount of heat energy that has been absorbed by the material.

thermohaline circulation: A slow-moving, density-driven circulation of ocean water that runs nearly pole to pole and throughout the world's oceans. The current runs in a loop, both along the ocean surface and thousands of meters below it.

thermosphere: The highest level of the atmosphere.

topographic map: A two-dimensional map that shows elevation by using contour lines. See also *contour line; bathymetric map.*

tornado warning: A communication to the public that a tornado has been seen by someone or detected by radar.

tornado watch: A communication to the public that tornadoes are possible. In other words, thunderstorms with high winds and rain that may produce a tornado are in the area.

trade winds: Global winds that flow toward the equator, turning west as they go. See also *easterlies; westerlies.*

tropical depression: A low-pressure system of thunderstorms over warm ocean waters; it has no eye and has maximum winds of 61 kph, but may strengthen into a tropical storm or a hurricane.

tropical storm: An organized, low-pressure system of storms over warm ocean waters. It has wind speeds between 63 kph and 117 kph, and may rotate.

troposphere: The layer of the atmosphere closest to the earth, in which air moves in all directions. It is where most of the earth's weather takes place.

typhoon: A massive rotating storm that forms north of the equator in the western Pacific Ocean. It is equivalent to a hurricane in the Atlantic Ocean, eastern Pacific Ocean, or Caribbean Sea. See also *cyclone; hurricane.*

upwelling: The rising of cold, deep water from an ocean bottom.

vortex: The movement of liquids or gases in a spiral around a central axis. In a storm, it is the calm center area around which clouds spiral.

warm front: A boundary that occurs when a moving, warm air mass overrides a cold air mass ahead of it. See also *weather front.*

water cycle: The movement and exchange of water between the earth's land, atmosphere, and oceans.

waterspout: A rotating column of air over a large body of water.

water vapor: Water that has evaporated into a gas.

weather: The state of the atmosphere at a particular time and place. See also *climate.*

weather front: A boundary that forms when air masses meet that have different temperature, pressure, and humidity conditions. See also *cold front; occluded front; stationary front; warm front.*

weather satellite: An instrument that orbits the earth, taking photographs and collecting measurements.

westerlies: Global winds that flow from the west to the east. See also *easterlies; trade winds.*

Index

Photo Credits

Front Cover
NASA image courtesy of David Long, Brigham Young University, on the QuikSCAT Science Team, and the Jet Propulsion Laboratory

Lessons
2 NOOA/Collection of Wayne and Nancy Weikel, FEMA Fisheries Coordinators **4** NASA **6** NASA Goddard Space Flight Center Scientific Visualization Studio/Blue Marble data courtesy of Reto Stockli (NASA/GSFC) **9** U.S. Geological Survey/photo by Ilsa B. Kuffner **10 (top)** U.S. Geological Survey/ photo by Jessica K. Robertson **(bottom)** NOAA **12** Jennifer Heldmann/National Science Foundation **13** Library of Congress, Prints & Photographs Division, PAN US GEOG – California no. 269 **14 (top right)** NASA **(bottom left)** Image courtesy of Kurt Severance, NASA Langley Research Center **15** NASA **16** Jeff Schmaltz, NASA Visible Earth **17 (left)** NASA Jet Propulsion Laboratory/NASA Visible Earth **(right)** Images produced by Hal Pierce. Caption by Steve Lang and Hal Pierce/NASA Visible Earth **18 (top)** Courtesy of SOHO consortium. SOHO is a project of international cooperation between ESA and NASA. **(bottom)** Jeff Schmaltz, NASA Visible Earth **20** National Oceanic and Atmospheric Administration (NOAA)/Department of Commerce **22 (left)** NASA Goddard Space Flight Center/NOAA **(right)** NASA Goddard Space Flight Center/NOAA **28** NASA **30 (top left)** NOAA Department of Commerce/NOAA Photo Library, NOAA Central Library; OAR/ ERL/National Severe Storms Laboratory

(NSSL) **(bottom left)** NOAA Department of Commerce/OAR/ERL/National Severe Storms Laboratory (NSSL) **(right)** NOAA Department of Commerce **31** Ralph F. Kresge/NOAA/Department of Commerce **32** FEMA/Tim Burkitt **34** U.S. Agency for International Development (USAID) **38** daveynin/creativecommons.org **40** FEMA/ Patsy Lynch **42** ©2011 Carolina Biological Supply Company **53** Library of Congress, Prints & Photographs Division, LC-BH824-4499 **54 (top)** NSRC/Anne Williams **(bottom)** Smithsonian Institution Archives. Image SIA2012-7651. **55** Smithsonian Institution Archives. Image 84-2074. **58** Nick Hewson/creativecommons.org **60** NASA/ Goddard Space Flight Center Scientific Visualization Studio. The Next Generation Blue Marble data is courtesy of Reto Stockli (NASA/GSFC). MODIS data courtesy of Jeff Schmaltz, MODIS Rapid Response Project (NASA/GSFC). **68** Bob Ryan, Chief Meteorologist NBC 4 (WRC), Washington, D.C. **69** NOAA/Department of Commerce/ NOAA Photo Library, NOAA Central Library; OAR/ERL/National Severe Storms Laboratory (NSSL) **71 (top)** Bob Ryan, Chief Meteorologist NBC 4 (WRC), Washington, D.C. **(bottom)** Bob Ryan, Chief Meteorologist NBC 4 (WRC), Washington, D.C. **74** Fiona Shields/creativecommons.org **76** NASA image courtesy Jeff Schmaltz, MODIS Land Rapid Response Team at NASA GSFC **83** Hafiz Issadeen/creativecommons.org **85** NASA **88** FEMA/George Armstrong **90 (top)** FEMA/George Armstrong **(bottom)** FEMA/ George Armstrong **91** FEMA/Win Henderson

LC-DIG-ppmsca-09851 **(right)** U.S. Navy file photo **176** NOAA Rice Library of the National Centers for Coastal Ocean Science **177** Mykl Roventine/creativecommons.org **178** U.S. Navy photo by Mass Communication Specialist 3rd Class Kyle D. Gahlau **180** Image courtesy MODIS Ocean Group, NASA GSFC, and the University of Miami **182 (top)** NASA images by Robert Simmon based on Landsat-7 data **(bottom)** Sea Surface Temperature data from the Advanced Microwave Radiometer for EOS (AMSR-E), courtesy Remote Sensing Systems/NASA Visible Earth **183 (top)** U.S. Geological Survey/photo by Christopher Arp **(bottom)** NASA/Goddard Space Flight Center Scientific Visualization Studio. Blue Marble data is courtesy of Reto Stockli (NASA/GSFC) **186** NASA **187** NASA **188** NASA/Kurt Severance and Tim Marvel **190** NASA **191** NASA **192** William Larned/U.S. Fish and Wildlife Service **194** Based on graphic by NOAA **195** Randolph Femmer/life.nbii.gov **196** George Gentry/U.S. Fish and Wildlife Service **197** George Gentry/U.S. Fish and Wildlife Service **200** Library of Congress, Prints & Photographs Division, LC-DIG-ppmsca-22947 **201 (top)** Robert H. Mohlenbrock @ USDA-NRCS PLANTS Database / USDA NRCS. 1995. Northeast wetland flora: Field office guide to plant species. Northeast National Technical Center, Chester. **(bottom)** National Park Service **202 (top left)** Scott Wing, National Museum of Natural History, Smithsonian Institution **(bottom right)** Scott Wing, National Museum of Natural History, Smithsonian Institution **204** Angela Rucker/U. S. Agency for International Development **205** NASA Goddard Space Flight Center **206** Andrea Schaffer/creativecommons.org **208** NASA Goddard Space Flight Center **210** U.S. Navy photo by Boatswain's Mate 1st Class Christopher Dallaglio **212** Elizabeth Labunski/U.S. Fish and Wildlife Service **218** Kristan Hutchison/National Science Foundation **219** NASA Earth Observatory/Robert Simmon, Data Sources: NASA Goddard Institute for Space Studies, NOAA National Climatic Data Center, Met Office Hadley Centre/Climatic Research Unit, and the Japanese Meteorological Agency. **220** NASA image by Robert Simmon, based on GISS surface temperature analysis data **223** www.cgpgrey.com/creativecommons.org **224** Photo by Jurvetson (flickr) **226** U.S. Navy photo by Mass Communication Specialist 3rd Class Shawn J. Stewart **228** Terry McCrea/Smithsonian Institution